INFORMATION
OPERATIONS
MATTERS

Also by Leigh Armistead

Information Operations:
Warfare and the Hard Reality of Soft Power

Information Warfare:
Separating Hype from Reality

INFORMATION
OPERATIONS
MATTERS

BEST PRACTICES

LEIGH ARMISTEAD

POTOMAC BOOKS, INC.
WASHINGTON, D.C.

Library of Congress Cataloging-in-Publication Data
Armistead, Leigh, 1962–
  Information operations matters : best practices / Leigh Armistead. — 1st ed.
    p. cm.
  Includes bibliographical references and index.
  ISBN 978-1-59797-436-3 (pbk. : alk. paper)
  1. Information warfare—United States. 2. Network-centric operations (Military science)—United States. 3. United States—Military policy. I. Title.
  U163.A698 2010
  355.3'43—dc22

                                                    2010007938

Printed in the United States of America on acid-free paper that meets the American National Standards Institute Z39-48 Standard.

Potomac Books, Inc.
22841 Quicksilver Drive
Dulles, Virginia 20166

First Edition

10 9 8 7 6 5 4 3 2 1

Contents

Introduction

In 1998 the U.S. Department of Defense (DOD) released the first of a series of seminal policies on information operations (IO). Titled Joint Publication 3-13, it laid out for the first time in an unclassified format how the U.S. military forces could utilize this particular element of power. As a relatively newly defined activity, this publication proposed to revolutionize the manner in which warfare, diplomacy, business, and a number of other areas are conducted. However, this transformation in the U.S. government with regard to IO has not occurred over the last decade, and a significant gap exists in the capability of the federal bureaucracy to support operations in this arena. While strategic policy and doctrine have been developed and promulgated, primarily by the DOD, the actual conduct of IO activities and campaigns across the United States are normally performed at a much more tactical level. This gap between theory and reality exists because the interagency organizations are often unwilling or unable to make the transformational changes that are needed to best utilize information as an element of power. In this book, the author has developed definitions and models that articulate why this gap exists, as well as details specific strategies for utilizing IO in a manner that best optimizes its inherent capabilities. Specific recommendations are noted below and will be laid out in greater detail:

- Develop an academic theoretical construct for IO.
- Understand that different approaches and processes are needed to support IO.
- Establish an international IO standards effort.
- Meet IO training needs.

This book is more than just a reflection on the shifting nature of power. As the title *Information Operations* suggests, information is changing in this new era, and how a nation or federal agency understands that fact will greatly affect its ability to manipulate power to its advantage. Thus the overall goal of this book is to better demonstrate this revolution in power by bringing together these disparate themes and the different threads of information to show the tremendous changes that are occurring today. Relatively recent key sources were drawn on to show how the gaps in theory are perhaps one reason for the dichotomy that currently exists in IO. Likewise, the author also attempted to review the broad spectrum of published works on IO that has become available over the last decade. The goal of this review is to give a complete assessment of what needs to be done in the federal bureaucracy in order to continue moving forward with the development of IO. Feedback from the book participants and the literature review also indicated that there were a number of areas that were considered deficient with regard to the conduct of IO within the U.S. government. In addition, there is a series of common themes from both the literature review and book participant interviews that center on a few key points—the desire for strategic goals; the use of standards as well as integrated communication systems; tools; IO metrics; and the need for common training efforts to conduct IO activities across all federal agencies.

Yet there was also a dichotomy between the stated desires of the book participants and the desires of prominent IO authors and theorists concerning the published theory on the subject of the capabilities and the actual tactical reality of IO in the United States. This gap is the crux of this book and can be seen most clearly in the sections on conceptual models, which emphasized a desire by many of the book participants for a more comprehensive series of strategic IO efforts by the federal government in order to truly maximize the power inherent in IO. At the same time, there was also a realization among many of the participants of this project that these actions would not happen on a timely basis; instead, a more realistic approach involving a broader set of criteria and using a bottom-up methodology was considered more feasible. Likewise, a key concept that arose in the conduct of this research was the understanding that the road to success involves an actual limiting of the stated objectives into a useable set of concepts, definitions, theories, and capabilities that are attainable and feasible with the resources available to the federal bureaucracy. This departure from the early IO rhetoric to a more pragmatic approach is probably one of the most important items to take away from this tome. Many of the book participants have come to understand that they need to adjust their goals in order to succeed in IO. The original theory and policy was too radical and has since been modified as the federal agencies in the United States better understand what is truly needed to best utilize this new capability.

CHAPTER 1

UNDERSTANDING THE PROBLEM

The Next World War: Computers are the Weapons, and
the Front Line is Everywhere . . .
—*James Adams*

The contemporary world is transforming itself into the Information Age, which has been called an "era of networks" (Copeland 2000; Arquilla and Ronfeldt 1996, 2001). Loudly proclaimed by many throughout the world as a revolutionary process, it is interesting to compare and contrast the differences between rhetoric and reality, especially in the employment of IO. This transformation of traditional uses of power promises to revolutionize the manner in which warfare, diplomacy, business, and a number of other areas are conducted. However, the gap between proposed capability and actual conduct of operations in the U.S government is wide. Strategic doctrine and guidance to best utilize the power of information exist, but actual information campaigns are almost always conducted at a tactical level. The ideas in this book were taken from one hundred background interviews of practicing, mid-level officials of the interagency organizations that are involved in conducting information campaigns in the United States. It is hoped that the conclusions developed in this project will be useful for future IO planners and senior-level decision makers. This research was based on the following hypothesis and will go through a rigorous, theoretical methodology to develop a coherent set of findings as part of this effort.

Hypothesis: In the U.S government, a significant gap exists between the theory and reality of IO. The gap exists because the federal bureaucracy is unwilling or unable to make the transformational changes that are needed to best utilize information as an element of power.

Information Operations

Information as an element of power is, and always has been, a nebulous term, but in this new era it possesses a capability that is considered crucial to the success of U.S. national security. However, the method that best utilizes this element of power to support the requirements of the U.S. government is still unknown. This is because IO crosses so many boundaries within the interagency process that it is often very difficult to quantify exactly what constitutes an information campaign. One reason for this is that there are other organizations within the federal bureaucracy, such as the State Department (DOS), that have traditionally concentrated on diplomatic efforts to support U.S. interests abroad but are now being asked to facilitate strategic IO activities around the world instead. This kind of tasking is abnormal for these different cabinet agencies, and it also belies their normal chains of communication and day-to-day procedures. As a result, the most recent attempts to conduct strategic, high-level IO activities in the United States are often aborted in favor of a more tactical set of options that are normally conducted by the DOD as part of its standard set of operations. A good example of these dichotomies in IO is seen in three military activities conducted by U.S. forces over the last decade. Whether it was Kosovo, Afghanistan, or Iraq, the primary focus of these campaigns from the viewpoint of Washington, D.C., was on the military victory. In none of these operations did IO play the transformational force that its advocates have predicted. While a number of IO capabilities and related activities have been successfully utilized, these efforts are still almost solely concentrated at the tactical level. The strategic revolution in warfare advocated by informational power enthusiasts in the mid-1990s has not materialized as predicted and desired.

Understanding the power of information is more important than ever because the contemporary world is now witness to onslaughts of manipulated images, where nations, groups, and individuals attempt to manage the messages that they receive. Information campaigns have been advocated and theorized to be conducted in a very similar manner, whether one is selling merchandise, like a soft drink, or a threat to national security, like weapons of mass destruction. Acolytes of informational power advocate the idea that one needs to influence the mind of the consumer or the public to get them to believe in a product or cause. Early IO enthusiasts contend that it is all the same in this new era, where the nature of power has radically changed and perception management and computer network operations figure prominently as new capabilities. Yet to date, the book participants have not witnessed this transformation of the structure of power around the world. In particular, the changes envisioned within the U.S. government, particularly with respect to influence campaigns, have not yet occurred. In many cases, foreign policy operations are still conducted using traditional military and diplomatic methods.

Emerging IO Theory

However, notwithstanding these issues, the transformational ideas inherent in IO are crucial and must become a reliable capability of the American arsenal. As the events of September 11, 2001, indicate, military, political, or economic powers are often ineffective in dealing with these new kinds of threats to U.S. national security. The aforementioned terrorist attacks were a blow to the images of the American public and government that affected the perceptions of many people in this country. Only a comprehensive plan, in which information is a key element, can defeat the fear produced by the terrorist attacks. In this new era, all factors of power must be utilized, for as some academics argue, it will be networks that will be fighting networks in the future (Arquilla and Ronfeldt 1999). Good examples of this abound in Operation Enduring Freedom and Operation Iraqi Freedom, where networks in the form of information campaigns fight networks made up of perceptions. The side that will ultimately emerge from this epic conflict as the victor is the one that can best shape and influence the minds of their adversary and their allies (Advisory Group on Public Diplomacy for the Arab and Muslim World 2003).

Unfortunately, the shift from the Industrial Age to the information environment may not mean that the United States will remain the dominant player in the political arena. John Arquilla and David Ronfeldt also write that nation-states are losing power to hybrid structures within this interconnected architecture in which access and connectivity, including bandwidth, will be the two key pillars of any new organization. They posit that truth and guarded openness are the recommended approaches to be used in both the private and government sectors to conduct business. In their opinion, time zones will be more important than borders. It will be an age of small groups that use networks to conduct "swarming" attacks that will force changes in policy (Arquilla and Ronfeldt 1997a). Key features include:

- There are open communication links where speed is everything.
- There is little to no censorship, and the individual controls his own information flow.
- Truth and quality will surface, but not initially.
- There are weakening nation-states and strengthening networks (Arquilla and Ronfeldt, 1997b).

The changes that are mentioned in Arquilla and Ronfeldt's book *The Emergence of Noopolitik: Toward an American Information Strategy* are truly revolutionary and describe a profound shift in the nature of power. Unfortunately, this transformation has not been translated from a strategic concept to tactical actions (Kuusisto 2004). Thus the intent of this book is to fill that void and describe why the early strategic theory on IO does not match the current tactical reality.

The day-to-day reality of how IO is conducted by the United States

While much of the early policy concerning IO stated the need for a more strategic, centralized philosophy of executing a top-down process by the U.S. government, the day-to-day reality of operations is much different (CJCS MOP 30 1993; DODD S3600.1 1996; DOD JCS JP 3-13 1998). This early concentration on the development of high-level IO strategy mirrors the philosophy of a doctrine from an earlier era of the nation's history. During the Cold War, the United States and its allies and the Communist bloc were in a psychological confrontation between two competing, and essentially incompatible, ideologies emanating from Washington, D.C., and Moscow (Taylor, 2002). The Soviet Union and the Warsaw Pact were easily the most recognizable of the threats to the free world, but other nations, such as China, North Korea, Iran, Iraq, Syria, and Libya, were also part of the equation. This bipolar Cold War era was an arena of realist conflicts, such as Vietnam and Korea, with states acting as the prime actors and anarchy being a central theme. This was a war in the sense that nations were mobilized and armed forces were always at the ready to commit at a moment's notice if needed. A sense of urgency existed; high-level doctrine and strategy were developed to meet these perceived needs. However, it was not the military or diplomatic efforts that succeeded in ending this effort; instead, it was the economic and informational might of America that eventually prevailed. Today, the former Soviet Union is a shadow of its past existence, with a population lower than that of the United States, and it has had difficulty deploying a number of its forces in Kosovo because of equipment failures (R. Clarke 2000).

Why in this post–Cold War era, when the greatest threats known to mankind, such as a major, surprise nuclear attack are lessened, is the United States still under attack from a number of different enemies, including the Al Qaeda terrorist network? Once again, there are many reasons, but primarily it is because the perception of the enemy has changed. The lack of equilibrium during the Cold War, unrest in the Middle East, and conflict in Southwest Asia are all significant factors in this change. While there are still "rogue states" (in U.S. terms) that can occupy the politicians and give credence to budget appropriations, other groups, including extremist religious factions, are freer to operate and carry out attacks on the United States in this post–bipolar period. Most of these non-governmental organizations (NGOs) or terrorist groups are no longer operating underneath the umbrella of a superpower and therefore have much more autonomy than ever before. Over the past fifteen years, and especially in the last decade, there has been an explosion of attacks on the United States. In some of them, information has played a key role. While a number of these incidents were conducted by lone individuals, others were the work of activists, foreign military units, terrorists, and even nation-states. Thus we are now at a point where each of these terrorist operations highlight the vulnerability of America and its population to these new types of warfare, where information and the integration of the government play a key role. Yet as mentioned previously, there is a tremendous gap between the theoretical potential of IO and its day-to-day implementation, and there are many times when the U.S. government has difficulty defending itself from informational attacks in this new environment. Finally, there is also new technology in this environment that is supporting the traditional cultural and economic issues of third-world communities, which have given these populations a much greater power in this new dynamic.

The relevance of IO in the broader strategic theory

It is this concept of power and control of information that is the core of this research. In this book, most of the analysis focuses on the key areas of perception management and computer network operations within IO. The former is often referred to by different names, depending on which branch of the U.S. government is being referenced: psychological operations (DOD), public diplomacy (DOS), strategic communications (National Security Council (NSC)), or influence operations (White House)—all of these terms can be considered analogous, and the author has elected to use these terms somewhat interchangeably. Likewise, computer network operations can also go by different names, such as information assurance, computer security, cyber warfare, and computer network attack. Once again, the author has elected to use these terms somewhat interchangeably. In addition, while there are many other capabilities of IO that could also be examined, such as deception and electronic warfare, the author chose to narrow the book's subject to the two key areas mentioned earlier—the perception management and computer network operations components of IO. This is because it is the attempts to conduct these specific kinds of influence campaigns that the U.S. government has had the most difficulties with recently. These two areas are where the gap is the greatest between theory and reality. It is hoped that recommendations from this book will offer the most potential for change within the federal bureaucracy.

In this research, the use of a modified soft systems methodology approach and active interviews was deemed most appropriate as part of a qualitative procedure. In order to get the trust from this large group of government and academic participants, the author interviewed some of them repeatedly over a five-year period in what has been labeled as developing a sustained and intensive experience (Creswell 2003). Out of the original 100 background and book interview sessions with 63 different people, a total of 40 participants were ultimately selected so that the author could obtain the most current and accurate information about the current and future state of IO within the federal government and understand the nuances of the problem. Therefore, it is the intent of this book to answer not only research, but also to gather and collect the opinions of these key individuals as to what the federal bureaucracy should do in the future in order to better utilize this element of power. This book was designed to flesh out a different perspective of the U.S. government and to specifically examine the policy, personnel, and organizational modifications that are ongoing within these agencies as they attempt to transform themselves. Therefore, not only are the research interviews seeking to answer the "what is" question from these surveys and subsequent analysis, but they are also attempting to answer the "what should be" question.

In summary, this book is an attempt to review all of the disparate efforts by the various components of the federal bureaucracy to utilize different portions or capabilities of IO. In addition, this research has also attempted to investigate how the key agencies of the U.S. government can use the inherent power of information to better conduct influence or strategic communication campaigns in the future. Likewise, this book attempts to use these conceptual models to describe a strategy to best utilize IO by the United States. It is hoped that the outcome of this book will provide a process that can be used to transform these organizations in a manner that will better allow them to understand and use the power of information to meet the threats in the future. The bottom line is the question of whether the federal bureaucracy can conduct an effective information campaign in this changing environment while assuring the security of their networks and information systems in this new architecture. To do this, the United States may need to change its collective interagency structure that has evolved over the last two hundred years into a more networked organization that can master the issues in the Information Age. This is a crucial issue because it is uncertain whether America will remain a dominant player now that industrial capacity is not nearly as important to a nation as its interconnectivity of information nodes.

CHAPTER 2

A THEORETICAL REVIEW OF INFORMATION OPERATIONS IN THE UNITED STATES

In an age when terrorists move information at the speed of an e-mail, money at the speed of a wire transfer, and people at the speed of a commercial jetliner, the Defense Department is bogged down in the micromanagement and bureaucratic processes of the industrial age—not the information age. Some of our difficulties are self-imposed, to be sure. Some are the result of law and regulation. Together they have created a culture that too often stifles innovation. . . .

The point is this: we are fighting the first wars of the 21st century with a Defense Department that was fashioned to meet the challenges of the mid-20th century. We have an industrial age organization, yet we are living in an information age world, where new threats emerge suddenly, often without warning, to surprise us. We cannot afford not to change and rapidly, if we hope to live in that world (Rumsfeld 2003).

This quote by the former secretary of defense emphasizes the dichotomy that exists within the DOD today. The need for change is widely recognized across the bureaucracy, but implementation has been slow and uneven. Unfortunately, this condition is symptomatic of the federal bureaucracy as a whole. In this next section, the author outlines the development of IO in America as it has evolved over the last sixty years and compares it to the available literature in order to develop a cogent and coherent argument to understand the context of this research. While these publications are very diverse and range over many academic subject areas, including power, information, international relations, computer security, and organizational theory, each will be linked to the evolution of IO within the U.S. government to provide an understanding of their context. The reason for this diversity of literature is the incredibly extensive nature of IO itself. Because the definition of IO is so broad and generic, at once it is everything as well as nothing, it is very difficult to understand where to frame the boundaries of the discussion. There is no clear line or easy demarcation to determine what is or is not a part of IO; more often than not, the researcher is forced to cast the net far and wide in search of primary sources that reference this emerging capability. Therefore, the reader will notice a wide variety of sources as the author describes the development of this new warfare capability in the United States.

IO is a formal attempt by the U.S. government to develop a set of doctrinal approaches for its military and diplomatic forces to use and operationalize the power of information. Per the original primary DOD policy on IO, the target is the adversary decision makers. Therefore, the primacy of effort will be to coerce that person, or group of people, into doing or not doing a certain action (Joint Publication 3-13 1998). To affect the adversary decision maker, IO attempts to use many different capabilities, such as deception, psychological operations, and electronic warfare, to shape and influence the information environment. This is a very high-level and strategic approach to policy within the U.S. government, but as mentioned in chapter 1, "Understanding the Problem," IO is more often than not performed at a lower, tactical level. Therefore, this section aims at studying the available academic literature to evaluate the differences as IO has evolved into a full-fledged warfare area.

Theoretical Constructs

One of the challenges in this research is that it does not propose to update, challenge, adapt, or confront any of the traditional theories of international relations. This is because the changes described in this book represent a profound shift in the nature of power. Specifically, they discuss the huge transformation regarding power and information that has not been fully accepted by academics educated in the traditional theoretical schools. In addition, there has not been a vigorous debate within the scholarly journals on these precepts. In general, there tends to be a shortage of ideas and thoughts to adequately express these new ideas. "A new lexicon is needed for this purpose . . . there is a huge gap between our sense of profound transformations and our ability to grasp them from a huge shortage of the tools needed . . . our vocabulary and conceptual equipment for understanding the emergent world lag well behind the changes themselves" (Rosenau 1998, 33). There is a dichotomy, and the need exists for a new theoretical construct, one that can better model and explain events in this new Information Age.

As most analysts realize, that quest for a new theory is normally unfulfilled because we often ask our models to do too much. To begin with, most theories do not predict; instead, they provide ideas of what events are likely or not likely to happen. What theory does is help to organize facts, identify variables, and determine which factors are the most important. The understanding that there is no comprehensive theory of international relations often can go a long way toward explaining how useful these conceptual models can be. There is not one set of assumptions or structure that will answer all political questions of our times; instead, there is theory that can provide a map of the landscape. The field of international relations is about perception; the insight that an adversary or ally may have often comes from observation of the different forms of power that a state or group may have. Therefore, it was not surprising in the past that since military power was often the easiest factor to measure, this element of power tended to be given the greatest weight in any sort of calculation. But as history has demonstrated on numerous occasions, the ownership of a preponderance of military does not always translate into victory. The fungibility of military power, as expressed in Robert Keohane and Joseph Nye's theory on complex interdependence, is not nearly as high as many analysts believe, thereby giving false illusions as to its usability (1989). Other types of power, including political, economic, social, religious, and informational, all play a role as well, yet because they are difficult to measure or calculate, their potential is often neglected or reduced in importance. Thus this book has developed a theoretical construct and proposed a methodology that provides a hypothesis, a point of departure, a construct, and a framework in which to more comfortably view the events as they occur.

The primary focus of this book will therefore be on policy, organization, and training with respect to information within the three main government agencies involved with foreign policy in the United States—the White House and the NSC, the DOD, and the DOS. The two key areas of IO that this book examines for development are computer network operations and perception management. As previously mentioned, these are the two warfare areas that have changed the most within the last decade. Computer network operations, as noted previously, is an umbrella term that encompasses a wide range of cyber-related activities. For the purpose of this book, CNO is divided into three parts:

1. Offensive—Computer Network Attack
2. Defensive—Computer Network Defense and/or Information Assistance
3. Support—Computer Network Exploitation

Each of these areas has a role to play in this new and exciting warfare arena. While many people have visions of precision accuracy and war without needless violence, others have a vision of a kinder and gentler form of warfare for man to evolve to. Yet as many officials within the federal bureaucracy have come to realize as part of this study, capability about computer network operations often does not equate to reality. Thus while computer network attack is a current capability of the United States, some would say that it is so limited by legal, political, and security constraints that it is virtually useless to the unified combatant commanders.

Perception management is the other key area of IO that has changed significantly over the last decade. Through the use of computers, telecommunications, video, the Internet, e-mail, and other technological advances, the ability to shape an image or conduct an influence campaign has increased greatly. Instances that are mentioned in this book include the use of a video camera by the Somali warlord Hussein Aideed in 1993, the denial of service attacks by the Electronic Disturbance Theater in 1998, and the timing of the second explosion at the World Trade Center in 2001. All of these events were perception management campaigns designed to manipulate public opinion. In each case, the tools used were different, but the goal was the same—to produce an effect, or a perception, in the mind of the target. As shown in this book, the ability of the U.S. government to affect this capability has also changed radically over the last decade.

Earlier in this book, Keohane and Nye set forth arguments in their seminal book, *Power and Interdependence*, which describe in detail how these academics portrayed the changing role of information with regard to the power capabilities within the world political structure (1989, 23). Also mentioned previously were Arquilla and Ronfeldt, who in a series of books culminating in their much heralded *The Emergence of Noopolitik: Toward an American Information Strategy*, recognized that we now live in the Information Age—an era of networks, interdependence, international organizations, and transnational activities (1992, 1993, 1996, 1997a, 1997b, 1999). This latter set of authors stated their belief that nation-states are losing power to hybrid structures such as NGOs and multinational corporations within this interconnected architecture. Access and connectivity, including bandwidth, are two key pillars of these new organizations. Truth and guarded openness will be the approach used by both the private and government sector to conduct business. Arquilla and Ronfeldt felt that time zones were more important than borders and foretold of an age of small groups that will use networks to conduct swarming attacks to force changes in policy.

International Relations Theories—How they compare to IO

The lack of a defined theoretical construct surrounding IO led the author to first examine the methodologies that serve as a foundation for the international relations field. Through the use of theories and models, academics in this area hope to better understand the complicated proceedings of world politics. Charles Kegley stated that the "theory of international relations needs to perform four principle tasks. It should describe, explain, predict, and prescribe" (1995, 8). In this section of the book, the three major international relations theories—liberal, realist, and alternative—are all examined in detail to determine how well they can explain the changes brought on by this new element of power. All authors and theories are reviewed with respect to the four fundamental points outlined below:

1. Object of analysis and scope of enquiry
2. Purpose of social and political enquiry
3. Appropriate methodology
4. Whether international relations are distinct from, or related to, other fields (Burchill and Linklater 1996)

The "science" and theoretical constructs that comprise international relations are relatively new. The field was not separated from the larger domain of history until 1919. Much of the outgrowth of international relations can be attributed to the academic reaction to the horrors of World War I. A need was felt to study the lessons learned from this conflict in an attempt to try to prevent a war of this magnitude from happening again. Thus the majority of the effort in the interwar period was conducted by scholars from the United States and the United Kingdom to answering the following three questions:

1. What had war achieved, other than death and misery for millions?
2. Were there lessons from the war that could be learned to prevent a recurrence of conflict on this scale?

3. Was the war caused by mistake, misunderstanding, or malicious intent? (Burchill and Linklater 1996, 5)

LIBERALISM

Liberalism was the first of these academic theories to evolve within the new field of international relations. It grew as a reaction to the grim reality of World War I. By definition, a liberal view of international relations believes human nature is essentially good and altruistic. There is a prevalent, fundamental human concern for the welfare of others, and liberals believe that bad human behavior is not a product of evil people, but rather evil institutions. War is not considered inevitable, and liberals view war as an international problem, and that international society must eliminate anarchy by reorganizing itself (Kegley 1995, 4), and as is shown later, these beliefs will contrast sharply with those from the realism theory. There are many sub-categories within the liberal framework, and these include: international liberalism, liberal utopianism, neoliberalism, complex interdependence, and international regimes.

The object of analysis and scope of inquiry are probably the biggest differences between liberals and realists. A good example is the study of internal state politics as an explanation of a nation's actions. Some academics believe that it is precisely these internal politics that greatly affected the international economics. One researcher argued that domestic politics were the overriding concern of the majority of the policymakers and that any benefits associated with international policies were often outweighed by the high political price at home (Simmons 1994, 4–18). The study of human activity also seems to be a main focus of liberal research; whole books are devoted to the study of how nations begin wars. One liberal academic assumed that man is intelligent and is somehow trapped by his decisions. This led to a discussion of why man starts wars or refuses to get out of one when he knows better (Maoz 1990, xii). This paradox can be compared to those who attempt to analyze why there are cases of misperception in world politics. Any avenue of liberal beliefs is the study of the causes, consequences, perceptual errors, beliefs, and images that are used by decision makers (Jervis 1976, 3). These academics felt that the "perceptions of the world and of other actors diverge from reality in patterns that we can detect and for reasons that we can understand" (David 1991, 235). Thus one purpose of the liberal inquiry is to demonstrate how we can better understand man and the factors that affect his political decisions. The anarchy that is so prevalent in the realist theory is present within the state system, but it is not present at the international level. The states are not acting as independent units pursuing national interests, but rather as a vehicle for leaders' own personal gain (David 1991, 237).

If the methodology liberals use to study their craft is examined, most will agree that it is from a traditional viewpoint with an emphasis on history, law, and philosophy. A good example is analysis of international regimes, where one researcher uses the international aviation regime to compare and contrast the efficiency of different theories.

> The position of all regime theorists, regardless of whether they are institutionalists or modified structural realists, can be translated into a single hypothesis: Given the considerable interdependence in the world, which necessitates cooperation among states, international regimes are pervasive in the international system—particularly in issue areas that lie outside the zero-sum realm of security—and once created, they are likely to persist (Nayar 1995, 143).

Do liberals believe that the study of international relations is a separate and distinct academic field? To be effective, they must reach out to other domains and use research conducted in these disciplines. For example, Robert Jervis believes that psychologists' work with respect to international relations is important, but he is wary of applying it directly to case studies. However, Jervis also believes that "most international regimes scholars have paid no attention to psychology—that they have failed to recognize the importance of misperception" (1976, 6). Likewise, he also understands that if decision makers recognize the limitations of their mindsets, and if they attempt to try to see the world the way the other sees it, then they may be able to decrease the cases of misperception. Specifically, he suggested that to expose implicit assumptions and give a decision maker more freedom of choice, the formulation and application of alternative images should be encouraged. While this may be accomplished by the divergence of interests, goals, training, and information available within any large organization, oftentimes this is not enough. It is often difficult, both psychologically and politically, for any one person to examine many alternatives. Instead, Jervis suggests they should employ devil's advocates. There are limits to the utility of a devil's advocate who is not a true devil, but overall, Jervis believed that a minority view is needed to guard against cases of misperception. These devil's advocates can ensure that new information, rather than calling the established sub-goal into question, will not be interpreted within the old framework (Jervis 1976, 415, 416).

REALISM

Liberalism was the first major theory of international relations, and this was generally prevalent during the interwar period. It wasn't until the late 1930s, when the book *The Twenty Years' Crisis* was published, that realism was championed (Carr 1939). This theory was later refined to reflect the events of World War II and the Cold War (Morgenthau 1967; Waltz 1990). The realities of power politics during this period did much to cement the realist theory as the predominant school of thought within international relations for the next forty years. Major beliefs of this theory start with the idea that man is sinful and wicked by nature. He lusts for power and cannot eradicate this instinct. The struggle for power is an all-consuming goal, with all other interests subjugated. Therefore, nations will define the acquisition of power as in their best interest and will build military capabilities to maintain and defend themselves. The military will always be considered the primary source of power. States will not rely on allies to protect them, and treaties with other nations are only useful for balancing power. While these ideas do not constitute all of the concepts of realism, they give a broad view of the theory's basic assumptions. Sub-categories of realism include: neorealism, structural realism, international political economy, and decision-making theory.

As opposed to liberal theory, the realists are mainly focused on the international system and the nation-states in their research. This is evident in the book *Politics Among Nations: The Struggle for Power and Peace* by Hans Morgenthau, who many consider to be the first academic to advocate a realistic theory on international relations. Based on lectures given at the University of Chicago, Morgenthau tried to differentiate realist theory by listing its six principles and defining power, including the many elements and factors (1967). He also attempted to give realism a scientific approach and then took a very detailed analysis of the limitations of power and the problems in world politics. Morgenthau's opus is considered a magnificent attempt to produce a grand theory of international relations.

Realism as a theory has evolved greatly in its social and political inquiry from its initial development. The change was mainly an attempt to show how the anarchical nature of the international system is the overriding determinant on man. Kenneth Waltz took the themes from realism that had been espoused by E. H. Carr and Morganthau and refined them, developing the theory that is now known as neorealism. In his seminal work, Waltz used philosophers such as Saint Augustine, Thomas Hobbes, Immanuel Kant, Jean-Jacques Rousseau, and Baruch Spinoza to show that the root of all evil is man, and thus he is himself the root of the specific evil (1990, 3). He also quotes Rousseau to say that he finds the major causes of war neither in men nor in states but in the state system itself (1990, 11). These arguments and others are steps on the road to Waltz's theory that international relations are characterized by the absence of truly governmental institutions. It is this anarchy that forces states to act the way that they do. This is because "each state pursues its own interests, however defined, in ways it judges best. Force is a means of achieving the external ends of states because there exists no consistent, reliable process of reconciling the conflicts of interest that inevitably arise among similar units in a condition of anarchy"(Waltz 1990, 238). David Dessler, in his *International Organization* article "What's at Stake in the Agent-Structure Debate?" tries to take Waltz's theory one step further by developing a structural model of international relations. This transformational structural theory, Dessler argues, can better explain and develop decision-making processes, horizontal linkages, and a more comprehensive ontology (1989, 441, 474).

This development of theory is the heart of the debate between neorealists and neoliberals. Some of the most contentious ideas are not about theory so much as they are about the factors that define a theory. For example, power is a major focus in the study of realist theory. David Baldwin attempts to address these issues by analyzing what power is exactly and how it relates as a variable. He reviews much of the doctrine in this area, and his general consensus seems to be that power is not as well defined and useful as many people believe. He thinks the term is used too loosely and that there should be much more defining or narrowing of its use. Baldwin also thinks that the issue of fungibility is not nearly as great as many theorists would desire (1980, 161, 180).

The methodology used by realists can also be quite traditionalist. Keohane explores the growth of international organizations and their influence in the international regimes that has significantly changed the dynamic in the last decade. Some of this is owing to the extensive amount of international cooperation since World War II, although Keohane warns "a rising level of cooperation may be overwhelmed by discord, as increased interdependence and governmental intervention create more opportunities for policy conflict." He believes that international regimes "enhance the likelihood of cooperation by reducing the costs of making transactions that are consistent with the principles of the regimes. They increase the symmetry and improve the quality of the information that governments receive" (Keohane 1984, 3, 6, 244, 246).

The study of international regimes by realists is also important because it shows an evolving theological methodology to perhaps a closer relationship with liberalism. Nayar (1995, 168) shows this aspect in his article on aviation. Although primarily concerning international regimes, Nayar states that realism is more robust than previously given credit for. He believes that liberal institutionalism thinks of international regimes as representing shared values and norms of an evolving, if nascent, international community that transcends interstate conflict. Nayar then goes on to state that according to realism, international regimes are related to the interests and capabilities of states, and any cooperation among states is regarded as contingent and transient. Thus it is his belief that structural realism emerges with the superior explanatory power in the case of international regimes. Keohane has a similar argument that says that hegemony is not as important as cooperation, and that "cooperation is viewed by policymakers less as an end in itself than a means to a variety of other objectives" (1984, 10). He also states that while hegemony may be used to create cooperation, it is the willingness of governments to remain within international regimes long after they could have left that is similar to what Nayar argued in his article.

Realists also take a scientific approach to their study of theory methodologies. In Richard Ned Lebow and Janice Gross Stein's article in *World Politics*, the authors proceeded to denigrate many of the so-called tests and data that deterrence theorists had used. They questioned the validity and reliability of the data, the application of the deterrence definition, and how one can verify intent (1990, 347, 352). Realists also tend to address technology issues readily. In a 1990 *International Studies Quarterly* article, James Der Derian addressed some of the problems that operators are experiencing in conducting business in modern society. The speed at which decisions are made and information passed often overwhelms the policymaker. This model sounds similar to what Jervis was arguing about the rise of misperception by decision makers. Der Derian also believes "speed is the essence of war" and that time is more important than geography for success on the battlefield. In this article, Der Derian tries to bridge differences in theoretical approaches by arguing that the post–structural ideas of Keohane "can grasp—but never fully capture—the significance of these new forces for international relations" (Der Derian 1990, 307).

Thus the ideas of technology and the use of it in foreign policy is often crucial to realist mindset. They are also explicit in the development of IO theory. The factors that Der Derian discusses in his article—simulation, surveillance, and speed—can all be summarized by information technology. This is also the general consensus and thrust of the article by M. J. Shapiro (1990, 329, 339) in *International Studies Quarterly*. In this paper, Shapiro basically argues that foreign policy is no longer limited to diplomats and the government; instead, technology has made it available to the masses. There are more players involved, and they have a variety of interests and equities that must be met in order for an issue to be resolved. Some of these new players are multinational corporations, the media, and NGOs. In Shapiro's view, the masses are complicating the discourse of U.S. security policy. The media, in particular, gain Shapiro's ire; he believes that they have altered the ability of government officials to conduct foreign policy. This is very interesting because much of realists' consternation evolves from the fact that the nation-states are losing control in this new era. Politics are becoming more complicated because there are multiple players with different agendas that now all have access to the playing field as a result of the rise of information technology. These factors are exactly what gets advocates of IO so excited—namely that they demonstrate that the power of the government is being transferred to the people.

ALTERNATE INTERNATIONAL RELATIONS THEORIES

The final category of international relation theories reviewed as a possible construct for this book includes all of the so-called alternative issues. It has only been in the last two decades, since the fall of the Soviet Union in 1991, that a major challenge to the dominance of realism and liberalism has erupted within the international relations field. Some of these controversies were caused by the collapse of bipolarity, others by the perceived eroding stature of the nation-state. Whatever the reason, a whole host of alternative and competing theories have arisen that have challenged many sacred assumptions about international relations. These consist of Marxism, critical theory, feminist theory, ecological theory, post–modernism, institutionalism, and constructivism. As a result of their diverse backgrounds, there is no standard definition for alternative theories. Instead, advocates try to focus more on these types of alternative issues, bringing them out of the margins to ensure that their equities are adequately addressed.

The object of analysis in alternative international relations theory often addresses subjects that have been neglected by traditional international relations research. Likewise, their social and political interest areas tend to be vastly different than mainstream academics. This can be seen in Christine Sylvester's book, *Feminist Theory and International Relations in a Post Modern Era*. Sylvester argues that feminist theorizing would have affected all of the great international relations debates if women had been included. She also argues that this lack of feminist insight not only limits the effectiveness of these theories, it also shows that the international relations field is parochial in its scope of inquiry. In her book, Sylvester also attacks the "typical" methodology utilized by international relations academics as being too limited and not inclusive enough of all viewpoints (1994, 4).

Alternative theories, more than any other, tend to broaden the field of international relations. Martha Finnemore argues that scholars in the international relations field would do well to look into the academic work being conducted in sociology (1996). Although Jervis had argued that psychologists were limited in their ability to solve issues within international relations theory, Finnemore believes the opposite. She states that the institutionalist research conducted since the 1970s has done much to provide evidence of global cultural homogenization. The growing interdependence that she sees is a product of a "Westernization" of the world in which the notion of bureaucracies and markets are flourishing. In addition, because of the idea that a nation-state is the only legitimate unit that can operate in the world stage many areas are pushed into becoming a state when they are not equipped to do so (Finnemore 1996, 328, 336). Thus she argues that it is sociology's work on the individual and institutionalism that needs new emphasis in our modern society. Research conducted on the cultural awareness that an individual receives from the state is crucial to the development of the nation's identity. In her conclusion, she argues that by understanding the sociologist's research into the spreading of Western values, the international relations scholars may better understand some of the factors that they face around the world. In addition, Beth Simmons's work on international economics mentioned earlier also has important comments for adherents to the game theory model. Her research indicated that the internal political situation overwhelms any other thought processes, causing many extra conditions to be added on as new factors, making it is virtually impossible to try to compare and contrast equivalent behavior (1994, 283).

In conclusion, the value of a theory is based on its usefulness in adequately assessing world politics. With the dramatic events started by the end of the Cold War, the liberal ideology has regained much of its former status and has seriously challenged realism as the preeminent theory within the international relations field. This is not because the liberal theorists predicted all of the events of the preceding decade; it is mainly because realism as a theory did not! Likewise, the alternative theories may not represent a grand international relations theory, but they have chipped away at the importance of realism for not addressing the many factors that these advocates see as important modifiers. There are many ideas that influence political decisions, and all of these must be taken into account in forming a theory. For no matter what ideology or theory academics represent they still must comprehensively argue and ensure that their model can meet the four goals described in this paper. That is the basic question that every student must ask: is this theory relevant, and does it describe the events in adequate terms? If a theory cannot describe, explain, predict, and prescribe accurately world politics, is it really a theory at all?

Definitions of Power, Information, and IO

Traditional measures of military force, gross national product, population, energy, land, and minerals have continued to dominate discussions of the balance of power. These power resources still matter, and American leadership continues to depend on them as well as on the information edge. . . . . Information power is also hard to categorize because it cuts across all other military, economic, social, and political power resources, in some cases diminishing their strength, in others multiplying it (Nye and Owens 1996, 22).

After thoroughly reviewing the different academic theories that comprise the field of international relations, there were none in whole that matched the issues involved with regard to IO. The author conducted additional research in the areas of power and information, specifically involving the methodologies attributed to these issues. Power can mean many things to many people. Generally, its use is understood. It is clear who has power and who does not. Power is also one of those ubiquitous terms that everyone seems to understand but few can actually define. Morgenthau defined the elements of national power as geography, natural resources, industrial capacity, military preparedness, population, national character, national morale, and the quality of diplomacy and government (1967). Nowhere in Morgenthau's definition is the use of information seen as an element of power. So this begs the question—have the elements of power changed over the last four decades? A short answer is yes and no, depending on the sources that one reads. For example, in a recent study by RAND, a revised view of power was suggested that combined national resources and performance to create an updated version of military capability as shown in Figure 2.1.

Notice that technology is rated as the number one national resource, as opposed to the more traditional concepts Morgenthau listed that primarily involved physical assets. This is a huge change from older analyses, which concentrated much more on a mere "counting" of military assets and industrial plants. This RAND study goes straight to the concept that symbolizes the whole Information Age, namely, that the traditional power structure of the international community is being radically altered, thereby allowing nations, NGOs, small groups, and even individuals to gain an inordinate amount of power based solely on their information technology capability. The RAND researchers emphasize these ideas even more as they explore this concept further in the aforementioned study. This can be seen in Figure 2.2 where the critical areas of technology—the location of information and communications—are analyzed. This revised ordering of resources that comprise power is definitely a change from previous studies in which more traditional emphasis was placed on natural resources.

Not everyone agrees with these concepts; sometimes even people in the same research group don't agree. In another conference sponsored by RAND and the Central Intelligence Agency (CIA), analysts attempted to update the definitions and rankings of nations vis-à-vis power, and the main elements considered still consisted primarily of military and economic factors, such as gross domestic product. Technology was sometimes included in this study, but information as a separate and distinct stand-alone element of power was never elucidated (Treverton 2001, 17).

Although there is a general understanding that change is needed in this new information environment, at what rate or pace is not always agreed upon. There are many academics who advocate a more gradual view of the changing emphasis of power and information. For example, Massimo Tempestilli made the argument for the greater emphasis on the military uses of the informational element of power

Figure 2.1 – Views of Power (Source: Tellis et al., 2000, 8)

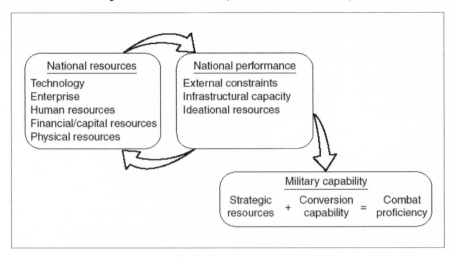

in his Master's thesis, *Waging Information Warfare: Making the Connection between Information and Power in a Transformed World* (1995). This is a slightly different slant than advocated by some academics who have called for a separate informational component or agency, similar to a cabinet agency, in the U.S. government. The U.S. government is organized with the cabinet agencies centered around each of the traditional respective areas of power—the DOD, the DOS, and Treasury or Commerce—with each having its own informational component. There are also interagency organizations, such as the NSC or National Economic Council, that still favor the concept of these three major elements of power (military, diplomatic, and economic). Yet a Department of Information still does not exist in the U.S. government because this form of power is still viewed by many as being very different from the more traditional elements. In fact, most of the participants in this book did not advocate a separate branch or cabinet agency for information. Likewise, Tempestilli argues that each of the three major elements of power—militarily, diplomatic, and economic—already have informational components and that the U.S. government does not need a new cabinet agency to focus solely on information. This horizontal integration of information versus a vertical division as its own element has both good and bad aspects from an IO policy perspective. Tempestilli argues, and the author agrees, that the cross fertilization between the informational components is better than a single, monolithic center for information. This concept follows a majority

Figure 2.2 – Revised Reordering of Technologies (Source: Tellis et al., 2000, 12)

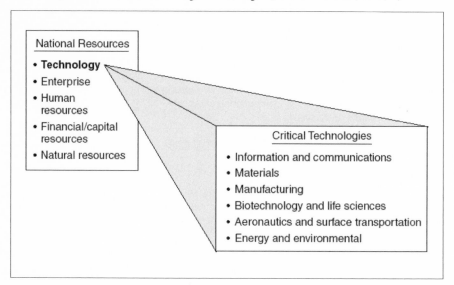

of the participants who advocated for a greater horizontal integration across the interagency spectrum. From a policy perspective, this can be seen in Joint Publication 3-13, *Information Operations*, that lists IO as an "Integrating Strategy"—that is, one that can bring together these disparate warfare areas (JP 3-13, 1998). Thus as Tempestilli originally advocated, and has been borne out in countless interviews, informational power cuts across the entire U.S. government structure and is not easily pigeonholed into a traditional cabinet structure. This is both a strength and weakness for understanding the power of information, because unlike military or diplomatic elements, it cannot be viewed in a traditional manner.

Therefore, it is not surprising that Tempestilli was only one of many authors who were commenting on the perceived notion of a revolution in military affairs that was occurring in the mid 1990s. There was a huge emphasis in the 1995–1996 timeframe, where a large number of articles by various authors highlighted the issues involved with the technological evolution of information. During this period, Eliot Cohen was one of these contributors who argued for a change in reorganization of the U.S. military to coordinate power in the Information Age. His concept was an attempt to solve the problems with incorporating information into the traditional, hierarchical government structure (1996). While his argument was not answered immediately, a number of changes have occurred over the last decade that completed the reorganization Cohen advocated. The U.S. government has significantly altered its organizational structure with respect to the functions of IO. However, as the participants of this book stated throughout the interview process, more still needs to be done to close the gap between strategic policy and the tactical reality of IO. IW is not the only revolution that has taken a long time to become established within the U.S. government; it took two decades after aviation was introduced into the military services during World War I for the government to realign and transform the military service into an organization that truly utilized the power of flight. IO will also take time and hard work, perhaps on a similar timeframe as aviation, for its potential to be truly realized.

CHANGING VIEWS OF POWER

The research in this book was conducted over the first decade after Joint Publications 3-13 was published, between 1999 and 2008. In this time, significant changes should have occurred with respect to these new views of power. If information is now accepted as an element of power, shouldn't there be dramatic changes from previous theories as well? Is the power of information new or different, as some advocates believe; or has information always been an element of power, but it could never be properly utilized? Said in another way, has information always been an element of power, and it is only now that technology can manage and harness that power? Critics of this new view of power have argued that because the world access to the Internet is not universal, this new technology cannot truly change global politics. Wriston notes that while this may be true, it is irrelevant (1997). The standard has been set, and the benchmark is high. These new views of information flow must be understood and respected. In fact, the percentage of overall access and connectivity to the Internet is on the verge of exploding as the combination of cellular technology and cheaper interface devices proliferate.

However, the question is whether access to technology necessarily equates to greater power to a group or nation. Once again, the short answer is—it depends. As Treverton relates in the report from RAND,

State power can be conceived at three levels: (1) resources or capabilities, or power-in-being; (2) how that power is converted through national processes; (3)

and power in outcomes, or which state prevails in particular circumstances. The starting point for thinking about—and developing metrics for—national power is to view states as "capability containers." Yet those capabilities—demographic, economic, technological, and the like—only become manifest through a process of conversion. States need to convert material resources into more usable instruments, such as combat proficiency. In the end, however, what policymakers care most about is not power as capability or power-in-being as converted through national ethos, politics, and social cohesion. They care about power in outcomes. That third level is by far the most elusive, for it is contingent and relative. It depends on power for what and against whom (2001, 11).

What is interesting about this third concept is that while it may be the most difficult to achieve, it may also offer the most promise. The four definitive bombing surveys of World War II and Vietnam prove that military power often does not automatically translate into desired outcomes. Mark Clodfelter, a retired air force officer, said as much in his book, *The Limits of Air Power, The American Bombing of North Vietnam* (1989). As the well-researched and documented official reports from the U.S. Air Force allude to, the massive bombing operations in that conflict often did not translate to shifts in the affected government or population's attitudes. As one veteran officer once quipped, "If the only tool you have is a hammer, every problem looks like a nail" (Hubbard 2004). Trying to take military power, in this case aviation assets, and translate it into recognizable outcomes is not an easy task.

The traditional central concepts of power in the form of national resources and the need to convert those resources into power and instruments of power are a key point of the last few pages as different academics have added and changed the common views of power. In addition, since IO as an academic study area crosses many issue lines, the development of suitable theoretical constructs with respect to power and information has not always been easy. A series of attempts that should be widely recognized can be attributed to Alvin Toffler and Heidi Toffler, two of the most prolific social authors with their three books *Future Shock* (1970), *The Third Wave* (1984), and *War and Anti-War* (1993). The influence of this couple and their publications on the DOD and U.S. government is so profound that they have had great effect on the general public and governments around the world. It is their futuristic forecasts of how we as a people are evolving with respect to the power of information that have made them famous. But they are not alone. Similar ideas about the changing of how the elements of power and information are viewed and used are shown in other literature as well. For example, Dan Kuehl from National Defense University (NDU) argues that information is an equal element of power just like its counterparts from the military, diplomatic, and economic realms (1997, 2004). In fact, in the U.S. military, the acronym DIME (Diplomatic, Information, Military, and Economic) is often used to express the concept that there are a number of elements that make up power and that the military aspect is not the only one that should be utilized.

SOFT POWER

The concept of soft power is not new. In her discussions on the idea of information as an element of power, Barbara Haskell was an early adopter of this philosophy. In an article on foreign policy, not only did she advocate the inclusion of information as an element of power, but she also included the social aspects of power (1980). Nye and others have brought forth the concept of soft power, which includes informational elements as well. In fact, Nye has continued an emphasis on this theme over the last eighteen years with a number of books and articles as the world has evolved in the post–Cold War era. Soft power is an interesting concept that basically argues that one can significantly influence other nations through the cultural and informational aspects of its society. As opposed to hard power with which analysts can often conduct intelligence to determine the potential of a perspective country, soft power is a more influential type of capability. It can be viewed as a theoretical construct that spans the gap between strategic policy and tactical operations of IO. In fact, soft power may be the capability that can attain that elusive outcome that was mentioned in an aforementioned RAND report (Nye 1990; Treverton 2001).

But what is soft power? It was originally defined by Nye as a concept that emphasizes the power of attraction, as opposed to the power of coercion. All forms of power are extremely hard to measure, and this is no exception. Some ideas that were put forward in another RAND study have attempted to develop metrics to measure power:

- Access to information—The government monopoly has eroded.
- Speed of reaction—Markets react in seconds, but governments are much slower, so the information technology revolution inevitably moved action away from governments toward nimbler organizations.

- New voices—The process created new channels of information and new, credible voices. The loudest voice, that of government, has become less dominant.
- Cheaper consultation—Because of nearly unlimited bandwidth, communication costs began to approach zero. Coordinating large and physically separated groups becomes much cheaper.
- Rapid change—Governments, by nature, are more likely to sustain the status quo than drive change, and so non-state actors are often the drivers by default.
- Changed boundaries in time and space—IT is driving the change, just as the invention of the printing press undermined the church's role as broker between people and their God (Treverton 2001, 13).

These are not the only metrics available, yet it is these outcomes that are most desired by government officials. Another recent series of reports by RAND on the "information revolution" illustrates the growing collection of data that is becoming available on information technology in particular. These include the number of Internet users, the Internet market size, and high-technology exports. So there are actually some metrics that are available, which is key because this kind of data may give researchers the ability to measure the factors that are needed to achieve outcomes without the use of military power. What is interesting about these concepts is that it is exactly these ideas that the three aforementioned authors advocate, which while radical in their time, have been generally accepted today. The problem is that there still remains a gap between the high-level strategic theory of IO and its day-to-day operations. The inability of the U.S. government to translate these lofty concepts—the actualization of the power of information—still remains elusive.

Chuck de Caro, a former Cable News Network reporter and Special Forces member, has taken this concept of soft power even further with his idea of "Soft-War." De Caro argues that conflict in the future can consist primarily of perception management campaigns with television as the primary medium. He believes that the vast majorities of populations get their informational news from television and that influence operations should be conducted using professionals from the entertainment industry (2003). This concept is probably one of the more coherent, cogent, and perhaps radical arguments that has evolved out of the IO debate. De Caro was interviewed on multiple occasions to expand on his ideas. It will be interesting to see how far he gets with these concepts.

Power and the Information Age

Can it be assumed from these books and articles that information to the masses translates directly into power? Perhaps. When Johannes Gutenberg developed the printing press, it vastly increased the ability of the average person to access the written word. Today, the media is much freer than in the past, but there are still many instances and locations where information is still rigidly controlled. The RAND studies on the information revolution around the world demonstrate these facts. Throughout this series, questions included: how has information technology changed political dynamics within the countries of a given region? And, how are the respective governments using information technology as a tool to govern? In their review, the authors from RAND analyzed from a largely bottom-up viewpoint the political dynamics of the actions and initiatives of citizens, civil society, NGOs, and political parties. These actions included the organization of protests of government policies and the overthrow of sitting regimes (Hachigian 2001, 55). The results of these surveys are fairly dramatic, revealing sharply rising access to information technology across a broad segment of the world's population.

Likewise, in understanding the dichotomy between the soft power and hard power aspects of IO, some books offer additional views on the power of information in the development of IO. For example, in *The Art of Information Warfare*, Richard Forno and Ronald Baklarz closely examine the perception management aspects of the power of information and specific deficiencies for the United States. These authors attempted use the writings of Sun Tzu as a model to relate to the different aspects of IO, and while they succeeded in some aspects, in others they were notably short, mainly because they did not address the computer networks operations aspects of information warfare. In addition, while Forno and Baklarz did address gaps for the U.S. government with respect to IO, they did not have realistic corrections or mitigations that could be used by the federal bureaucracy.

These changes combined with the capabilities of information technology and the role of the media with respect to the government show the dramatic changes that have occurred with respect to the elements of power in the sixty years since the end of World War II. In his article, "The Military/Media Clash and the New Principle of War, Media Spin," Marc Felman notes as much in describing the historical trends of relationships between these two entities and his belief that media pools were not the future answer to "handling" the press in combat (1993). Written after Operation

Desert Storm, it is interesting to see how the intervening decade leading up to Operation Iraqi Freedom allowed the military an even better understanding of the power of information, and this was reflected in the press coverage of the latter campaign. However, it is still very difficult to generalize how the control of the media reflects directly on this element of power. These changes are very interesting, because in two separate documents published by the DOD and DOS, it is readily apparent that both cabinet-level agencies are mutually coming to the understanding of the need for change in their informational policy in this new environment (Joint Publication 3-13 1998; U.S. Advisory Commission on Public Diplomacy 2000). In fact, both of these documents are excellent examples of the theoretical disconnect in IO between strategic policy and tactical reality because both documents are ahead of their time. Unfortunately, neither publication has fulfilled its original mandate or promise, but they have paved the way for additional intellectual discussion.

Likewise, this 2000 publication by the DOS also echoes the changes reflected in academic journals and other military sources by emphasizing the need for a new style of diplomacy, one more akin to the revolution in military affairs. The areas that are highlighted in the publications listed above also continue to focus on new computer systems, notably information technology, that can be used to better aid the traditional diplomatic missions of the DOS. In addition, a shift from traditional secretive diplomacy to a more open public diplomacy role has been advocated as well, with calls for increases in financial resources and a reformation of the U.S. foreign affairs agencies (Brookings Institution 1997). These were not the only sources that highlight the importance of the management of information as a source of power. Another good example is Phillip Taylor's *British Propaganda in the Twentieth Century: Selling Democracy*, which is perhaps the only book that has been dedicated to analyzing the power of information with respect to propaganda, public diplomacy, psychological operations, and deception. Taylor ties together these disparate and obscure missions in an attempt to understand the role of perception management in the twentieth century (Taylor 1999, 2002).

These views on the changing role of power in the Information Age are also reflected in other publications. Jamie Metzl wrote an article that shows the mindset of senior officials in the Clinton administration as far as the potential of perception management, in particular public diplomacy and international public information with respect the changing role of power. Titled "Network Diplomacy," the article was published in the *Georgetown Journal of International Affairs* (2001). What is especially interesting is that much of this article was written while Metzl was serving in key roles with the development of Presidential Decision Directive (PDD) 68, "International Public Information," at both the NSC and the DOS. Interviewed over a four-year period (1999–2003), Metzl was very insightful in his comments about how the U.S. government bureaucracy attempted to come to grips with the power of information through the implementation of new policy (2000). Yet the inability of the White House (both the Clinton and Bush administrations) to follow through on any strategic communications, public diplomacy, or international public information effort on a long-term basis—including PDD 68—is crucial to the arguments of this book. While it is notable and laudable that all these documents are released and that a tremendous amount of effort has gone into their research and publication, progress in making these changes has been very slow. It will take a long time to fully realize the true capabilities of IO, and the gap between IO theory and reality may continue to exist for the foreseeable future.

Taken together, all of these references help to define the changing evolution of power. This is advantageous because it gives a baseline from which to understand the new roles of information. As a reference term, information can be very perplexing. It can be technically oriented to mean the packets of data on the Internet and a piece of electronic bandwidth, or it may be more socially oriented to mean human-to-human contact, but information is really much more than that. In reality, information can be perceived as the glue that binds the power process together. Without it, there can be no international systematic structure. Therefore, just like power, information also has many meanings to different people. This book uses the following definition articulated by the U.S. military in the Joint Publication 3-13 (1998, 131):

1. Facts, data, or instructions in any medium or form
2. The meaning that a human assigns to data by means of the known conventions used in their representation

Information is more than just a definition. As an admirer of the concept of the information society has stated, "Information exists. It does not need to be perceived to exist. It does not need to be understood to exist. It requires no intelligence to interpret it. It does not have to have meaning to exist. It simply exists" (Webster 1995,

27). It is concepts such as this that can make it difficult for the layman to understand how information can be a source of power. Formerly, the control of information could be somewhat restricted to official government channels. However, this is no longer the case, not only because of the changes in the computer and telecommunications industries, but also because of the interconnectivity of the world. For example, some analysts, such as John Brown, believe that the power of information has now shifted to the masses and away from the government. "Compared with 10 years ago, the world is a more seamless informational space, as we move from a world of distinct national informational spaces into a more trans-national informational sphere." (Brown 2002, 4). However, there are others who believe that nothing radical has changed with respect to information and power. This dichotomy was extremely apparent to the author in the research interviews and produced the two opposing conceptual models for the use of IO in the U.S. government.

CHAPTER 3

THE DEVELOPMENT OF IO

From a theoretical and strategic IO perspective, Joint Publication 3-13 was the seminal document for the U.S. military and other organizations across the federal bureaucracy (Brown 2002). For the first time in an unclassified format, the DOD issued the definitive concept of how America plans to conduct operations in the Information Age. This pamphlet showed how this particular element of power can be used to affect the world politic and just how important the Joint Chiefs of Staff consider it to be. In effect, Joint Publication 3-13 defined the strategic vision of what IO truly could do for the U.S. government. But there have been issues and disconnects in this DOD policy from its inception in 1998; a number of attempts have been made to rewrite or update this doctrine to make it more user friendly. As this book was being finished, a new update to this policy had recently been released to accommodate all of the changes that are occurring within the military with respect to IO. The updated IO policy is more constrained and resembles the original, narrower command and control warfare definition as defined later in this chapter. The new IO policy has also tried to substantially narrow the theoretical gap that exists today. But it is uncertain whether this is a step back in IO theory or if is it more of an admission of the reality of how the United States can realistically conduct operations in this warfare area. While there is no definitive answer, the author believes that, based on the data collected throughout this interview process, it is probably the latter.

The Role of Information in Warfare

There are a number of publications that have appeared over the last decade that seem to address the role of information in this new environment. For example,

Ryan Henry and C. Edward Peartree argue for a new political theory based on the power of information (1998). Participants in this research agree with this need. In fact, the search for a suitable theoretical construct was a long and involved process because of the diverse and complicated nature of information. This is not to say that political theorists have not tried to develop new theoretical constructs with respect to IO. For example, as mentioned earlier, RAND has been very active in writing proposals for new informational policy for this era. The first of these was Arquilla and Ronfeldt's *In Athena's Camp*, which was quickly followed by Roger Molander's *Strategic Information Warfare Rising*. In both of these books, the authors argue for a policy shift with an emphasis on the national or strategic level of war, where information should leverage the most power (Arquilla and Ronfeldt 1997a; Molander 1998). Interestingly, it is this call for strategic IO actions, and the subsequent lack of follow-on examples, that is the crux of this research; namely, there is a gap between tactical IO activities and strategic policy. A third book published by RAND during this same time period, *The Changing Role of Information in Warfare* (Khalilzad and White 1999) also offers a strategic promise of the utilization of this newfound power. But once again, there is little follow-on progress from the U.S. government, as the DOD did not make the wholesale changes as proposed in that book. Instead, what has happened over the last decade has been a series of small, discrete steps to slowly grow the DOD's capability with respect to IO, all of which will be laid out later in this chapter.

The next publication by RAND, *The Emergence of Noopolitik: Toward an American Information Strategy* (Arquilla and Ronfeldt 1999), is especially interesting because the authors have attempted to redefine political theory with respect to international relations. They tried to develop a new international relations strategy based on the power of information, in essence trying to answer Henry and Peartree's call for a new political theory for the power of information in the process. Likewise, Arquilla and Ronfeldt also argued that there is a gap between perception management and computer network operations in IO that has not been adequately addressed by the U.S. government, and that more strategic analysis must be conducted. These authors believe that an overall IO policy must be pulled together from disparate pieces in order to build a doctrine that can be analyzed as a coherent whole. To quote the authors, "Strategy, at its best, knits together ends and means, no matter how various or disparate, into a cohesive pattern (1999, 5)." As Arquilla and Ronfeldt also stated, they believe that these two ideational poles encompassed by manage-

ment and computer network operations are the keys to developing an overarching
IO theory. In addition, they believe that in order for their new theory to succeed,
a strategic analysis or linkage should be developed between these often disparate
and insular communities (1999, 3). As will be seen later in this section, attempts to
build this overarching strategy continue to fall short. Likewise, the linkages between
the different portions are not nearly as strong as Arquilla or Ronfeldt advocated.
Unfortunately, it also appears that the U.S. government still fails to realize much
of the promise of the power inherent in IO, as advocated by Arquilla and Ronfeldt.

The Role of Information in Government Organizations

Other academics have also tried to analyze these changes. RAND embarked on a
three-year-long study of the effects of the information revolution on governmental
organizations, and key discussion areas included the political, governmental, business, financial, social, and cultural dimensions. RAND analysts noted such changes
occurred for two general reasons:

1. Traditional mechanisms of governance (for example, taxation, regulation, and
    licensing) are becoming increasingly problematic as the information revolution
    allows action beyond the reach of national governments.
2. The distribution of political power is changing as new non-state actors are being
    empowered by the information revolution in the business, social, and political
    realms at the subnational, transnational, and supranational levels.

These academics believe that governments will have to find mechanisms to deal
with these changes and new actors, and different nations often take different approaches. How this is accomplished will define the roles of power and information in
the nation-state, and especially in the United States. Other RAND publications about
information have followed as well, including a study in 2000 titled *Information and
Biological Revolutions: Global Governance Challenges* (Fukuyama and Wagner 2000).
This text examines the new elements confronting political leaders in the post–Cold
War era and offers suggestions for change. An additional study by Marcin Libicki on
the governance and development of the global information grid was published that
same year. In an analogy to this book, Libicki debates whether the U.S. Air Force
should adopt a top-down, centralized approach to management of these services, or
whether a more decentralized, bottom-up approach would be more successful. In

this particular case, Libicki believes that it is inappropriate for the military branch to develop an enterprise-wide management control—a top-down approach—at this time (2000). This series of thoughts were very similar to data derived from other book participants, but there is still a disconnect in all of these RAND studies because they fail to acknowledge the large gap between their proposed strategic doctrines of IO and the day-to-day reality of tactical operations. There is also a serious disconnect between what many academics believe is possible to do with respect to IO and what the U.S. government is willing to do in practice. This gap still exists today as evidenced by the data gained from the participants in this project.

However, RAND was not the only semi-government agency interested in the power of information at this time. The Armed Forces Communications and Electronics Association (AFCEA) was also busy publishing a series on information warfare over a four-year period. Their titles include:

- *Cyberwar: Security, Strategy, and Conflict in the Information Age* (Campen, Dearth, and Goodden 1996)
- *Cyberwar 2.0: Myths, Mysteries, and Reality* (Campen and Dearth 1998)
- *Cyberwar 3.0: Human Factors in Information Operations and Future Conflict* (Campen and Dearth 2000)

Written as an anthology, these books emphasize the evolution of the strategic and theoretical analysis of the warfare area since 1996. It is interesting to notice how the AFCEA also stressed perception management and computer network operations, the same key areas that Arquilla and Ronfeldt focused on. Two of the editors, Douglas Dearth and Dan Kuehl, were interviewed on multiple occasions for this research, and they contributed valuable insight into the changing role of information with respect to power in the U.S. government. Unfortunately, this series was discontinued after the third edition, and no further books are likely to be published. Offering an opportunity for twenty to twenty-five respected practitioners of the tradecraft to update the general public and academic community on IO activities, this series has been sorely missed. It was in the original *Cyberwar* book where much of the hope and promise that constituted the revolution in military affairs movement of the mid-1990s can be traced back to. Overall, the contributors of this series appear to be generally optimistic about the future of information warfare, but there were also cautionary tales, especially with reference to the threat of cyber war. However, this was a series of seminal publications, a set of ideas that framed much of the discussion for IO when it was only starting to be recognized as a unique warfare area. The editors of the original *Cyberwar* book were also fortunate to be able to include an introduction by Thomas Rona, the original creator of the term information warfare. Developed two decades earlier, he recognized the value of information and data within the context of nuclear war and the bipolar threat that existed at that time. Rona tied in the threats to the civilian infrastructure from IO, which was quite unique and led to a nice dialogue among the disparate commentators in this series. He understood that changes in information brought threats to not only the warriors in the field, but to civilians and society as well.

Therefore, from these books, articles, and interviews, it can be understood that there are many factors in the equation of power with respect to the changing role of information. Some of these scholars believe that information is now the most important element of power because it is the most fungible of the different fundamentals of influence, which would relate to the fundamental shift from the Industrial Age to the Information Age. Likewise, some of the authors mentioned are concerned that the rise of information as an element of power is diminishing other facets and concepts of power, such as sovereignty. In addition, as noted by Richard Rosecrance in *The Rise of the Virtual State*, the fungibility inherent in information gives the average citizen much more power than he had in the Industrial Age (1999). Information that was previously only accessible to the rich and powerful is now widely available. Communication around the world has increased so much that country-to-country dialogue is not solely limited to diplomats; instead, it is conducted through millions of other conduits. These immense changes allude to the difficulty that countries, including the United States, face in this new environment. Arquilla and Ronfeldt also discuss these same issues, most specifically the role of information in the conduct of foreign policy, as another element of power in conjunction with military, diplomatic, and economic elements. These authors also acknowledge that it is exactly this ability to manipulate and manage the power of information that makes concepts like IO so useful, and so destabilizing, to the status quo. For what all of these academics understand and relate in their publications is how in today's environment, groups, organizations, nation-states, and even individuals can now influence policy at the systemic level simply by using information. This was not necessarily the case during the Cold War, but the vast explosion in technology, particularly in telecommunications and media propagation over the last fifteen plus years, has forever changed the control over this power paradigm (Arquilla and Ronfeldt 1997a).

This change and recognition that informational power in the form of IO is changing the way that the United States conducts its military and foreign policy initiatives can also be seen in articles and books other than the official publications already mentioned (Joint Publication 3-13 1998; U.S. Advisory Commission 2000). Barry Fulton stated as much and describes how the DOS must change to adapt to the influx of informational power (1998). For probably more than any governmental bureaucracy, the DOS had a near monopoly on control of communication between governmental leaders, but with the advent of the Internet, twenty-four-hour news channels, satellite television, and worldwide newspapers, that is no longer the case. Unfortunately, few, if any, of Fulton's suggestions were followed through. In 2000 the U.S. Advisory Commission on Public Diplomacy published a similar study titled *A New Diplomacy for the Information Age*. Little was done to change this federal agency, and the DOS continues to grapple with these changes. Few, if any, of these studies or critical recommendations for changing this cabinet agency have been implemented. But what is also interesting is that these studies all mirror the DOD's voluminous output of publications during the same time period. For instance, Fulton's tome was, in effect, a corollary to Joint Publication 3-13. The Joint Publication will be explained in the next section, but both of these documents were attempts by their respective federal organizations to come to grips with the power of information and incorporate it into their processes and methodologies.

There are obviously other writers on this subject and a large number of books have been published in the last decade about IO, computer warfare, cyber security, and net war. For example, the book, *The Next World War: Computers are the Weapons and the Front Line is Everywhere* (Adams 1998), forecasted a multitude of changes in the information world owing to the increased connectivity of the globe. While James Adams did not emphasize globalization as much as connectivity, there is clearly a linkage between the two. This link is shown by Stephen Flanagan, Ellen Frost, and Richard Kugler in their NDU series on globalization and national security titled *Challenges of the Global Century: Report of the Project on Globalization and National Security*. This two-volume set features fifty chapters on a far ranging set of topics, including strategic implications, emerging priorities for the United States, and the challenges ahead, including both global and regional trends (Flanagan et al. 2001). Both of these books emphasize the new roles of information around the world and how it is changing the dynamics of power. In addition, this eighteen-month project confirmed some of the key themes with respect to the changing role of power and information as they affect the U.S. government, including the impact of the media and a bifurcated world order.

Other authors and social scientists have also examined the effects of information as it has impacted the U.S. government. They have come to their own interesting conclusions. James Gleick, a journalist and author of the books *Faster* and *What Just Happened: A Chronicle from the Information Frontier*, offers a unique perspective on the evolution of the information society and its cumulative changes to people and the way that they live in the American culture. Blending science and cultural journalism, Gleick offers different perspectives on the effects of the increased information flow and how it is speeding up aspects of life in the United States. Unfortunately, Gleick is only of an observer of the changes brought on by the Information Age, and he offers no concrete solutions for improvement by the U.S. government (Gleick 1999, 2002). Howard Rheingold is similar. He has written two books, *The Virtual Community* and *Smart Mobs: The Next Social Revolution*. Both of these publications comment on the incredible changes in society around the world as a result of information technology and how this newfound power has empowered these citizens to conduct new initiatives (Rheingold 2000, 2003). Once again, there is little useful advice or recommendations for changes on how to best utilize IO in this new environment. However, what these two authors have done is cobble together disparate ideas that range across a wide spectrum of the information environment and bring them together in one place that is commercially available to all. It is these kinds of books that senior-level government leadership should read in order to get a feel for how fast the world is changing around them. In addition, these real-world examples are incredibly useful in helping explain the paradigm shift that is occurring with the information revolution.

Understanding Information Operations

Even with the publication of all of these books and articles, IO is still not understood very well. To many people, IO is simply computer warfare, yet it is about much more than that. In the United States, it is an attempt by the federal bureaucracy to develop a strategy to use all of the capabilities of information to affect the many issues that it deals with in the post–Cold War era. With these changes in the elements of power has come the realization that militarily, the United States could not solve all of its problems through kinetic means. Therefore, IO is an attempt to bring these different facets of power to bear on an adversary in a synergistic manner to achieve our national objectives. For a long time, it was hard for the DOD to address or even intelligently discuss the concepts of IO. This was because there was no common directive or publication available. This led to questions and confusion regarding definitions and lexicon that could not be fully addressed until the release of a coherent strategic policy, which came in the form of the Joint Publication 3-13, *Joint Doctrine for Information Operations*, which was published in October 1998. For the first time, the DOD was distributing the doctrinal principles involved in conducting IO in an unclassified document—a key milestone in the development and use of IO within the U.S. government.

Per Joint Publication 3-13, the real key to making IO effective across the federal bureaucracy was to ensure that the horizontal integration and coordination of the interagency organizations were conducted early on while still in the peacetime environment. IO can be a very effective tool for shaping the environment in the pre–hostilities phase. At its most successful, it can avoid or minimize the actual need for hostilities. However, that is not always possible. There are still differences in definitions of IO. This is owing to the fact that while it was explicitly defined in 1998 by the DOD, the concepts of information warfare go back to two different DOD directives that were issued in 1993 and 1996. In addition, there are other subtle differences between these two warfare or mission areas; the primary doctrinal difference is that information warfare contains six elements and is mostly involved with the conduct of operations during actual combat. IO includes these six capabilities, but it also includes two activities that are occasionally integrated or related to information warfare—public affairs and civil affairs. Likewise, IO is not only broader than information warfare, it is also intended to be conducted as a strategic campaign throughout the full spectrum of conflict—from peace to war and back to peace—across the federal bureaucracy. Thus for all these reasons, IO is considered much more comprehensive than information warfare, and it is in IO that the full integration across government agencies and with private industry must occur (Joint Publication 3-13 1998).

Information Warfare	Information Operations	
ELEMENTS	CAPABILITIES	RELATED ACTIVITIES
Computer Network Attack	Computer Network Attack	Public Affairs
Deception	Deception	Civil Affairs
Destruction	Destruction	
Electronic Warfare	Electronic Warfare	
Operations Security	Operations Security	
Psychological Operations	Psychological Operations	

The elements, capabilities, and related activities of information warfare and IO are separate and distinct warfare elements. Most have very old traditions and long-standing histories that do not necessarily mean that every action conducted in these areas is always associated with IO. There are elements of destruction that are not part of an IO campaign; likewise, not every public affairs activity has to be tied to information operations. In reality, all elements and their components of national power should be integrated into a satisfactorily planned, designed, and executed information strategy. If this is not done, the United States may not attain its national security goals in the new millennium.

Thus the concept of IO is intended to use these different capabilities and related activities to produce effects in an integrated fashion. Therefore, while one can try to use all eight capabilities and related activities to conduct an operation, a good IO plan will probably only incorporate a few of these warfare areas (Giessler 2004). The basic idea is that kinetic means are not always necessary, and instead, for IO to work properly, the operators must understand the environment, assess their interests and the adversary's pressure points, and then use whichever capability or related activity best affects the adversary. Thus IO is much more of an intensive study of not only an adversary, but also of personal forces, which is more than perhaps many current military commanders have grown accustomed to (Kuehl 2004).

IO Development in the United States

As mentioned earlier, the use of information to influence foreign audiences is not new. Throughout the twentieth century, the United States attempted to use information as a tool of public diplomacy to influence foreign audiences around the world. President Woodrow Wilson created the Committee on Public Information in 1917.

During World War II, President Franklin D. Roosevelt established the Office of War Information (OWI), which included the Voice of America (Campen and Dearth 2000; Armistead 2003). This agency and its overseas component were the forerunner of the U.S. Information Agency (USIA), which was the home to public diplomacy within the federal structure for almost fifty years. Defined as government activities intended to understand, inform, and influence foreign publics, public diplomacy is one of the forms of IO, along with perception management, strategic communications, and influence campaigns, that comprise the crux of this book. It was this strategic use of information that became a key factor of U.S. foreign policy in the Cold War, where information was disseminated to worldwide audiences by television and radio broadcasts in the form of a state-to-state dialogue. And the United States was not alone. Nations throughout history and to the present day have tried to use information to influence other countries as well as their own citizens. How successful they were in those attempts often depended on a number of factors, including cultural and psychological biases, as well as their means and methods of technology used to transmit that information.

These ideas are not new. The books *Science of Coercion* (Simpson 1994) and *Psychological Operations and Political Warfare in Long-Term Strategic Planning* (Radvanyi 1990) are only two of the more prominent publications that offer detailed academic examinations of the psychological operations portion of IO. Considered the most difficult to execute, and perhaps hardest to understand, it is perception management, part of the softer side of IO, which seems to need the most future research as indicated by the various authors of these tomes. But there is a big difference between offering future research topics and concrete solutions to ensuring the tactical reality of IO as a mission area. So in many aspects, both of these texts fall short of offering a viable solution to incorporating IO into day-to-day military operations. This aspect was also mentioned by a multitude of the research interviewees who suggested that it was the softer side of IO where the greatest gap existed between strategic policy and tactical reality. So it is good to see that the DOD has come to the realization that its doctrine must more closely match its capabilities and is changing the perception management portions of IO policy in its most recent version of the new Joint Publication 3-13.

However, IO in the U.S. government is not just about perception management. Computer network operations play a major role as well. With the tremendous advances in computers and technology, the nature in which governments and countries interact has changed dramatically as well. A number of books have attempted to address these issues, some of which were written by authors previously mentioned. For example, Arquilla and Ronfeldt published the book *Networks and Netwar* in which they describe the future of terror, crime, and militancy. It follows the theme of their earlier work, *The Advent of Netwar*, with a collection of essays mainly written from a social netwar perspective by a distinguished collection of authors. A good update to their previous books, it is in this new publication that Arquilla and Ronfeldt were able to expand on the emphasis on the importance of the networks as an enabling framework for IO. Luckily, the editors were also able to add an afterword following the events of 9/11 to tie together their themes.

Likewise, Owens and Offley presented their ideas on ensuring the adequacy of U.S. military power through more cooperative uses of information, joint operations, and more emphasis on flexibility by the respective services in their book, *Lifting the Fog of War*. While not a pure IO book, Owens and Offley did tie in some of the strategic concepts associated with this warfare area and highlighted the follow-on efforts from the earlier revolution in military training efforts that are especially illuminating from an IO perspective (2001). Other authors have also attempted to write about the development of IO within the U.S. government as well, including David Alberts, the author of *The Information Age* (1997). It is part of a series funded by the DOD that fostered a debate and helped to frame the course of changes within the U.S. military. In this series, a number of strategies are debated about what the right course is to ensure that America has the best military for future operations and how IO plays a major role in much of this debate. The overall consensus from this discussion is that the strategy of IO must be very closely related to actual conduct of warfare (Alberts 2002). Wayne Hall generally agrees with both of these assessments in his book *Stray Voltage: War in the Information Age*. He also notes that there is a significant disconnect with regard to the maturation of strategic IO within the federal bureaucracy (2003). From a series of tours within the DOD and the U.S. government, Hall is knowledgeable enough to understand that this new warfare area cuts across many operational boundaries. He knows that it is not just enough to concentrate on the technical aspects of IO to be successful, but that the softer aspects must be understood as well.

Moving on to other strategic aspects of IO policy within the U.S. government, a number of authors have written books on the technical aspects—the harder side of IO. Dorothy Denning, formerly of Georgetown University and now with the Naval Postgraduate School, linked computer network operations with IO in her book *Information Warfare and Security*. In this seminal publication, she addressed a number of information security concerns, including information assurance, and was one of the first authors to lay out these key aspects of IO in an unclassified forum. Two other publications that were mentioned earlier, *The Information Revolution and National Security* and *Strategic Warfare in Cyberspace* are similar to Denning's work in that the authors also compared the development of current national strategy with efforts to coordinate cyber policy and offered recommendations for the future. All three of these books analyzed the links between power, information, doctrine, and security policy, and are good ties between the harder and softer aspects of IO. In addition, these books are also important because it was during this period in which the DOD was laying out the current policy on IO in the form of Joint Publication 3-13 as well as creating the key IO organizations that are described later in this chapter.

On a final note, a significant development for IO from a theoretical standpoint was a doctoral book by Myriam Dunn that was published by the Swiss Federal Institute of Technology in Zurich. Titled *Information Age Conflicts: a Study of the Information Revolution and a Changing Operating Environment*, the primary difference from this book in Dunn's approach was the use of structural realism as her theoretical construct. Dunn confronted the dilemma of the inconsistencies in her theory by trying to build a model that delineated key challenges associated with the Information Age. She examined all of the traditional international relations theories and dismissed them as inadequate to truly explain the changing environment. She admitted that her proposed choice of structural realism had major flaws in its use as a tool for modeling the power of information, and thus she was restricted in her ability to adequately explain IO. In addition, she also understood the constraints of all forms of realism, which maintain the state as the primary actor (2002). As noted earlier, because of the radical changes in the power structure within this new era, the state is no longer the primary player in the Information Age. It was for this reason and others that will be delineated in later chapters that a soft systems methodology was used as the theoretical construct for this book instead of an international relations type of theory. It shows how the power of information is radically changing the traditional power structure around the world.

Historical IO Case Studies

FROM HIROSHIMA TO THE BERLIN WALL—THE COLD WAR ERA

As explained earlier, IO is not new; there are great examples of the different parts and capabilities of IO within the U.S. government. Historical data abounds on the capabilities over the last sixty years, with a good illustration from the immediate post–World War II era. In this specific case, the Truman administration wanted to strengthen and coordinate the foreign information measures in order to attain U.S. national objectives, specifically from a perception management perspective. In an attempt to stop the spreading specter of communism, the NSC passed an executive directive, NSC 4, "Coordination of Foreign Information Measures," on December 17, 1947. This policy was expressly designed to combat the extensive propaganda campaign being conducted by the Soviet Union at the time. Written to exploit and promote the message of economic aid that the United States was delivering to a number of foreign nations, especially in Europe, this policy directive was also crucial in the interwar period, for there was no existing government agency specifically tasked to conduct strategic information campaigns for the American public. Therefore, this policy document was meant to serve as an interagency coordination mechanism, led by the Secretary of State.

Coinciding with these efforts by the NSC to develop organic information programs was a concern within Congress about the DOS's ability to propagandize U.S. citizens as well as foreign nationals. Therefore, new legislation was enacted to ensure a separation existed between these two capabilities. It is somewhat amazing in this era of a throwaway and disposable society that much of the government agencies are actually constrained by a law more than a half century old. In fact, the U.S. Information and Educational Exchange Act of 1948 specifically forbids the U.S. foreign policy apparatus from conducting propaganda on U.S. citizens. Much of this concern by Congress was in direct response to the period following World War II when the conduct of public affairs and psychological operations within the U.S. government security structure was unrestrained. There were operations conducted by the OWI and the U.S. Information Service that quickly raised questions about the propriety of these activities. Therefore, to ultimately coordinate the activities of the foreign affairs organizations, the U.S. Information and Educational Exchange Act was enacted as a sweeping legislative bill that forbade the DOS from conducting propaganda, psychological operations, or public affairs on the American public. Interestingly enough, it still stands and is a major restraining component on IO efforts within the U.S. government.

This new act created a serious dilemma for the DOS in 1948 because it created a dichotomy between policy and operations. The new Assistant Secretary of State was supposed to conduct public diplomacy with a target audience of foreign nationals abroad, yet was also supposed to manage a public relations campaign aimed for domestic consumption for the DOS. There is an extremely fine line between building information tools on the same subject for two different audiences (Bernhard 1997). To make matters worse, the Act made it illegal to conduct public diplomacy operations on the American people and directed that separate budgets exist for public diplomacy and public affairs. So not only did these staffs have to differentiate products between their different audiences, but they also needed to do so under separate operating authorities and budget tasking. In 1952 the USIA was created, and its sole purpose was to create a public diplomacy arm that could conduct these activities legally abroad against foreign citizens (Bernhard 1997).

Thus for almost fifty years, the USIA was the main organization responsible for the conduct of public diplomacy and information campaigns by the federal government. Formed under Reorganization Plan No. 8 of the Information and Educational Exchange Act, this new activity encompassed most of the information programs of the DOS at that time (Armistead 2003). When it was created, the lines of authority for this new agency were unique. It operated as an independent organization; its director reported to the president through the NSC, and the director coordinated his own separate budget. These factors and resentment of their freedom within the agency became major elements in later reorganization efforts by the DOS over the next five decades.

In addition, efforts were also underway during this period to strengthen the use of public diplomacy as a tool for the United States. On March 6, 1984, the Reagan administration published a policy titled National Security Decision Directive (NSDD)130, "U.S. International Information Policy." This document was envisioned to be a strategic instrument for shaping fundamental political and ideological trends around the world on a long-term basis and ultimately affecting the behavior of governments (Campen and Dearth 2000). Written by the staff of the President who was described as the "great communicator," it is not surprising that President Reagan would believe in the transformational power of information. Recognizing that a strong international interagency capability was needed, NSDD 130 was a successor to NSC Directive 4.

THE REVOLUTION IN MILITARY AFFAIRS AND THE GLOBAL
WAR ON TERRORISM

These changes during the Reagan administration were just the beginning of a maturation of IO policy in America. It was in the first Bush administration and the demise of the Soviet threat to the continental United States in 1989 that the greatest shift in policy with respect to IO began in the U.S. government. From the lessons learned during the experiences of the Cold War, it became clear to war strategists that the side that controlled the most information and retained the ability to accurately manipulate and conduct an influence campaign was going to be victorious (Owens and Offley 2001, 100). This was most apparent when, immediately after the fall of the Soviet Union, strategic planners at the Joint Chiefs of Staff began to think and write new strategy, most of which was highly classified, on the use of information as a war fighting tool. Owing to the restrictive nature of this new strategy, the first document, DOD Document TS3600.1, was kept at the Top Secret level throughout its use (TS3600.1 1992).

While this publication started a dialogue on information warfare within the DOD, its classification ultimately restrained a more general doctrinal exchange. Thus the need for a strategy to fit these revolutions in technology still existed, and a new concept termed command and control warfare was quickly developed. Officially released as a Chairman of the Joint Chiefs of Staff Memorandum of Policy 30, *Command and Control Warfare* (March 8, 1993), this document laid out for the first time in an unclassified format the interaction of these different informational disciplines, which when combined could give the war strategists the information warfare advantage (CJCS MOP 30 1993). As originally defined, command and control warfare contained the following five pillars:

1. Destruction
2. Deception
3. Psychological Operations
4. Operations Security
5. Electronic Warfare

Intelligence supported these five pillars in order to conduct both offensive and defensive aspects of this capability. While some quarters of the military greeted this new concept of warfare with enthusiasm, others were wary of any new doctrinal developments. However, the ability to integrate these different military disciplines to conduct nodal analysis against enemy command and control targets was also highly lauded as a great improvement (CJCS MOP 30 1993). Many units and all four military services in the United States developed command and control warfare cells and began training in this new doctrine throughout the mid-1990s. But there was a conflict between the Joint Staff and the Defense Secretariat doctrine, since information warfare was a much broader attempt to tackle the issue of information as a force multiplier, while command and control warfare was more narrowly defined to apply only to the five pillars mentioned above (CJCS MOP 30 1993; S3600.1 1996; JP 3-13 1998). The fact that the United States was writing strategy to conduct operations against nations in peacetime was considered very risky; therefore, official information warfare policy remained highly classified throughout much of the 1990s (Pilecki 2000).

The U.S. military also recognized the need to develop commands and agencies to conduct these types of warfare in the Information Age; therefore, even though doctrine was still in the formative stage, organizational changes began to occur in the early 1990s. The Joint Electronic Warfare Center at Kelly AFB in San Antonio, Texas, was renamed the Joint Command and Control Warfare Center in 1993, and would later be renamed the Joint Information Operations Center in October 1999, and finally the Joint Information Operations Warfare Command in 2004. The uniformed services also created a number of other new agencies beginning in 1995, including:

- U.S. Air Force—Air Force Information Warfare Center
- U.S. Army—Land Information Warfare Activity—later changed to the 1st Information Operations Command
- U.S. Navy—Fleet Information Warfare Center—renamed the Naval Information Operations Command and now subordinate to Navy Network Warfare Command

In addition to organizational changes by the services, new courses and schools were also being developed to teach new tactics. The NDU created the School of

Information Warfare and Strategy in 1994. It was a ten-month-long academic curriculum designed to immerse the National War College students in the academic theory of information warfare. Held for two years, the NDU graduated sixteen students the first year and thirty-two in the second. However, the course was subsequently cancelled in 1996. This may have been owing to a belief that information warfare instruction needed to be disseminated to a wider audience. Shorter courses and classes were developed to teach a larger audience of NDU students. These existed for several years, including a five-day intermediate information warfare course for mid-grade officers and a two-day information warfare overview for senior officers. All were cancelled by mid-2003 (Giessler 2004), yet IO was still taught at NDU as a series of embedded lectures in different curriculums. In 2008 there was also movement to reinstate IO as a major subject topic at this institute with the establishment of a Masters program. The other official DOD joint course on information warfare is also taught at NDU's Joint Forces Staff College, formerly the Armed Forces Staff College, in Norfolk, Virginia. Held for two weeks, seven times a year, the current Joint Information Operations course is aimed primarily at mid-grade officers or civilian equivalent government personnel, who are serving in an IO cell or billet with a joint agency. A planner's course that takes these students to the next level was also developed in 2001 and is still widely taught. As can be seen, doctrine continued to develop after the publication of the command and control warfare doctrine *Memorandum of Policy* (MOP) 30 in March 1993. The formation of information warfare agencies and commands in the 1995–1996 period filled voids in the services and helped resolve the conflict in the development of information doctrine and policy within the U.S. government.

Thus the formation of information warfare agencies and commands in the 1995–1996 time frame also helped to resolve the conflict in the development of IO doctrine and policy within the U.S. government. However, since DODD S3600.1 was still classified Secret, it limited greater discussion on the differences between IO and information warfare. But this constraint was somewhat muted in 1996 when the DOD presented Joint Vision 2010, a white paper written to establish a vision for how the U.S. military will operate in the uncertain future. For the first time in an unclassified format, IO was formally defined as "those actions taken to affect an adversary's information and information systems while defending one's own information and information systems" (Joint Chiefs of Staff 1996, 69). To implement this vision and achieve "full spectrum dominance," four operational concepts were introduced in this publication:

1. Dominant maneuver
2. Precision engagement
3. Full dimensional engagement
4. Focused logistics

The essential enabler for all four of these concepts was doctrinally encapsulated as information superiority (Joint Chiefs of Staff 1996, 69). Defined as "the capability to collect, process, and disseminate an uninterrupted flow of information, while exploiting or denying an adversary's ability to do the same," information superiority consists of three components of which information operations is a prime factor. In addition to these doctrinal changes, the period of the mid- and late-1990s was also a time of early experimentation. In the same time period, the aforementioned Joseph Nye and retired Admiral Bill Owens were also recognizing that the United States should take advantage of its information superiority in the post–Cold War era. They published an article in *Foreign Affairs* titled "America's Information Age." This piece corroborated much of what the DOD was attempting to do with its revolution in military affairs by describing the country's move into a "Third Wave," away from an industrial nation toward an informational society (Nye and Owens 1996). But once again the supporters for IO advocate developing a high-level strategic policy but provide little information on how to actually achieve these goals. In addition, as was noted in this article, with these perceived advantages came threats; America is most often recognized as the nation with the most vulnerability from a cyber attack. Other authors followed Nye and Owens's beliefs, advocating a radical change in the manner that the U.S. government could conduct warfare. However, the theoretical disconnect continued to exist, because the changes advocated by these authors were often too radical and too fast for the DOD to carry out.

For example, Winn Schwartau is probably at least as well known for his book on IO, *Information Warfare: Cyberterrorism—Protecting Your Personal Security in the Electronic Age*, as he is for his testimony before Congress and his annual IO conference, INFOWARCON. His efforts to heighten awareness about IO often seemed over the top, but he believed that he was successful in getting the American population to understand this new threat (Schwartau 1996, 2003). The problem that resulted from these methods was once again a disconnect between lofty promises of new and wonderful capabilities with respect to IO and the reality of what could actually be done, especially in the early stages of IO between 1995 and 2001. In the author's opinion, this sensationalism of IO in that time period actually did a disservice to the emerging warfare area, because it oversold the reality of what IO could really do for the U.S. government. Unfulfilled promises then led to dissatisfaction, which may have led to disbelief. Ultimately, hype had to be separated from reality in order to move ahead with the "real" capability of IO. The author believes that this has been accomplished over the last decade as IO has been operationalized and brought into the mainstream of DOD operations.

James Adams also oversold the problem in his book *The Next World War: Computers are the Weapons and the Front Line is Everywhere*. His publication was more of a reflection and observation of trends, much like Rheingold and Gleick's works, which are mentioned elsewhere in this chapter. But what Adams did do correctly was show that the traditional boundaries of warfare had been removed in the Information Age, and U.S. citizens could no longer count on the military to protect them (1998, 2000). While Adams was not the only author to understand this important point, his book was one of the more useful in explaining the consequences of this new environment. Likewise, the same can also be said of Arquilla and Ronfeldt's book *The Advent of Netwar*. One in a series of publications by these prolific authors on this topic, what distinguishes this book from their others is its emphasis on a new kind of warfare, one fought by networks against other networks (Arquilla and Ronfeldt 1996). Interestingly enough, this concept of network-centric warfare has also been adopted by the U.S. military, including the U.S. Navy. Network-centric warfare has been championed by retired Vice Adm. Arthur Cebrowski and has gone a long way toward operationalizing IO (Cebrowski 2003). This is very interesting because in this case, one actually had a concept for utilizing IO that was advocated and accepted by the DOD. Vice Admiral Cebrowski echoed many of the key points in these books when he discussed the huge changes that are occurring as power shifts from the Industrial Age to an Information Age.

Roger Molander, an early advocate of IO, expressed similar thoughts in his book, *Strategic Information Warfare Rising*, and during his interview with the author (1998, 2003). He understood that the threats in this new environment were not from traditional adversaries; instead, they were from a variety of organizations and entities that were not previously thought to possess this type of capability. While much of his book focused on the cyber threat, this author believes that the intervening years between the interview for this book and its publication have shifted some of Molander's view. In this time period, the events of 9/11 occurred, as did the invasion of Afghanistan and Iraq, all of which may have contributed to Molander's emphasis on the softer side of IO during his interview in 2003. It would have been interesting to see if the same shift occurred with Greg Rattray, the author of *Strategic Warfare in Cyberspace*, which was published by the Massachusetts Institute of Technology. Both his book and his interview for the research in this book were completed before the events of 9/11. Rattray's manuscript had a huge emphasis on cyber warfare, but the book may have been different if it had been published after the events of 9/11. Rattray, unlike virtually all of the other authors mentioned in this book, did get a chance to operationalize his theory when he was selected to be a commander for a U.S. Air Force IO squadron in San Antonio, Texas, in 2003.

TRANSLATING POWER INTO OUTCOMES—KOSOVO (1999)

By describing the successes and failures of the military operations in Kosovo, the next section demonstrates a good case study on the effects of perception management and the U.S. government. This case study is on a massive air campaign conducted in early 1999 by a coalition of United States and NATO air forces against the former Yugoslavia over its policies of genocide in the Serbian province of Kosovo. The allied coalition flew over thirty-four thousand combat sorties in a seventy-eight-day period of bombing, inflicting massive destruction on Serbia's economic infrastructure. Rather than bringing stability to the region as IO doctrine dictates, NATO's operation actually created greater regional instability and the potential for future conflicts.

The strategic bombing campaigns that were exercised by the Allies against Germany in World War II were first described by General Giulio Douhet, the renowned Italian air power theorist. Gen. Douhet envisioned a total warfare, where a nation's military, industry, and population were attacked to bring about a swift and total defeat. These campaigns were supposed to be a thing of the past for the United States. IO doctrine does not advocate attrition bombing attacks and wholesale destruction against an adversary. Indeed, the advent of precision-guided munitions and effects-based targeting has added a new dimension to using physical destruction as an information weapon. The mere ability to destroy one of an adversary's high-value targets while leaving the surrounding area virtually unscathed sends a very potent psychological message. First, it demonstrates the precision, lethality, and superiority of American weapons technology. More importantly from an IO perspective, limiting collateral damage and physical destruction gives the adversary less ammunition for hostile propaganda directed against the United States. Second, the U.S. military has now conditioned the international media to low collateral damage and precision engagement, so when the occasional accident occurs and a non-military target is hit, the media will tend to amplify the effects of the accident. By its sheer excellence, the recent U.S. aerial campaigns have inadvertently set an inescapable standard for minimizing collateral damage. However, there is much more to IO than just a targeting or destruction campaign.

Both the domestic and foreign public expect the United States to avoid inflicting massive collateral damage and civilian casualties since it has the technological means to do so. Failure to accomplish this strategy makes the United States a target of criticism by domestic and foreign media and politicians. The very manner in which the United States uses physical destruction may provide an information tool for an adversary. When the United States uses physical destruction to manipulate the behavior of an adversary, it must defend itself against the hostile propaganda of that adversary and strive to maintain absolute credibility. Therefore, it is critical that the public affairs and psychological operations messages describing the use of physical destruction be absolutely accurate. While sounding impressive, this lofty list of NATO's achievements later proved inaccurate. In what may have been an overzealous desire to demonstrate positive results from a two-month-long air campaign that was beginning to draw considerable international criticism, NATO put its credibility on the line with statements that the Serbian military knew to be inaccurate. Given that the National Army in Kosovo was the target of U.S. and international public information and psychological operations efforts, any loss of credibility with the target audience ultimately only harmed these operations.

Therefore, although the original premise from the allied leadership was that this would be a short strategic bombing operation, the war quickly began to drag on as the effects of the strikes did not faze the Serbians. In fact, it was not until almost eight weeks into this campaign that IO type strategies were developed to try to use new methods to bring pressure on Slobodan Milošević himself. The bombing didn't bring about the desired results, and other tactics were needed against the dictator. Some of these specific attempts to conduct an information campaign were aimed at discrediting his policies while undermining his economic means to continue the conduct of the war (Arkin 2001). To do this, high-level diplomats from the allied coalition conducted near-simultaneous press briefings emphasizing the fact that Serbia was condoning Milošević's genocide actions. In the meantime, bombing missions were conducted against specific factories and industries that were funding the upper leadership. Detailed and tailored messages were also sent to these same Serbian government officials in an attempt to influence them to shift away their allegiance from Milošević. All of these actions taken together, along with the military, diplomatic, and economic pressure, are what many people believe helped bring an end to this conflict. Many of the details are still classified, but reports are starting to leak out that it was the information campaign rather than the bombing campaign that was ultimately successful as a perception management tool to ultimately out Milošević from Serbia (Arkin 2001).

To summarize, Kosovo will probably rank as the second Information War. Through the use of advanced information dissemination, including faxes, e-mail, and webpages, as well as perception management campaigns, this conflict was fought for the hearts and minds of a worldwide audience. Where the ultimate changes were actually made was in the detailed, tailored targeting of the key individuals who could affect the decision makers. That is what was different about this operation and the use of information. It was, for the first time, a war in which information was recognized as the primary weapon that was used to bring about a decisive end to a conflict.

CHAPTER 4

RECENT CHANGES TO IO

Policy Changes: The Information Operations Roadmap

In this section, a very detailed review of the most recent changes in IO policy and organizations will be undertaken to compare to the recommendations that constitute the theoretical concepts developed in this research. Undoubtedly the most significant recent policy change that impacts IO from a U.S. standpoint was the publication of the *Information Operations Roadmap* (DOD 2003). This directive proposes a way ahead for the U.S. military forces with regard to the future of IO. The *2001 Quadrennial Defense Review* identified IO as one of six critical goals supporting DOD transformation, and it set forth the objective of making IO a "core capability" for future U.S. forces. The *IO Roadmap* further identified three critical areas in which U.S. capabilities must be improved. The first of these was an improved ability to "fight the net." This desire stemmed from the realization that in an era of network-centric warfare, protecting the networks on which the DOD depends is essential to U.S. military capability. The second of these critical areas was the need to improve psychological operations in the DOD. This translated into making it more integrated with, and supportive of, national-level themes and objectives and enhancing the ability of the United States to impact adversary decision making. Finally, the third crucial area was the need for U.S. forces to conduct offensive operations in, and via, the electromagnetic spectrum, including computer network attack and electronic warfare capabilities.

From these three critical areas, the *IO Roadmap* further recommended a series of actions to improve the overall offensive IO capabilities of the DOD. The first of these was to develop a common understanding of IO, and it offered a new definition of IO that would eventually be issued to the world via a revised Joint Publication 3-13 and a revised DODD 3600 (JP 3-13, 2006; DOD S3600, 2006). The *IO Roadmap* also stressed the need to both consolidate oversight and advocacy for IO while simultaneously delegating capabilities to the combatant commanders. To do this, the U.S. Strategic Command's (STRATCOM) IO role was expanded and strengthened until it became, in effect, "the IO command." This need to create a core of trained and educated IO personnel and the requirement to improve the ability to analyze IO operations and effects were both cited in the *IO Roadmap*'s recommendations. In addition, there were also suggestions for the improvement of each of the five core competencies of IO as defined by the *Roadmap*—namely, computer network operations (which includes attack, defense, and exploitation), electronic warfare, military deception, operations security, and psychological operations. The need to clarify the "lanes in the road" between psychological operations, public affairs, and public diplomacy was also emphasized. Finally, IO's place in the budget process needed increased transparency in order to clarify what resources IO actually had and what would be needed to provide a stronger, more robust and comprehensive set of capabilities. In all, the full *IO Roadmap* laid out fifty-seven specific recommendations designed to develop specific elements of the overall recommendations as discussed above.

The new definition of IO published in the *IO Roadmap* was also very much centered on the military aspects of information and was almost a verbatim return to that contained in the early 1990s doctrine for command and control warfare, defining IO as: "The integrated employment of the core capabilities of electronic warfare, computer network operations, psychological operations, military deception, and operations security, in concert with specified supporting and related capabilities, to influence, disrupt, corrupt or usurp adversarial human and automated decision-making while protecting our own" (DOD 2003, 11). This new definition was a significant narrowing of IO's scope from what had been laid out in the 1998 Joint Publication 3-13, which defined IO as "actions taken to affect adversary information and information systems while protecting our own" (JP 3-13 1998). That earlier approach in 1998 was much broader and more inclusive of other federal IO activities and tended to focus on effects rather than means. It was also more difficult to resource. Traditionally, the military services are responsible for organizing, training, and equipping forces, and they complained that nothing in the original 1998 definition could be directly tied to military programs. However, the new definition

(2006) could be immediately tied to several long-standing and well-integrated force programs. The new definition also included several controversial elements, most of which were related to the word "influence." The "lanes in the road" issue contained in the IO Road Map recommendations was the most controversial element because it brought together into one discussion several activities and communities that traditionally have viewed each other with great suspicion. The links and relationships between public affairs (whether DOD or DOS), psychological operations, military deception, and public diplomacy were undeniable in a theoretical sense, but in reality, turf battles, organizational cultures, and concerns over roles and responsibilities all intermixed to create an environment that often does not embrace changes easily.

Thus, the *IO Roadmap*'s definition of IO was actively formalized across the military services with the release of the newly revised Joint Publication 3-13 (JP 3-13 2006). The old Joint Publication 3-13 had been in effect for more than seven years, during which time much had changed in the IO environment. While the old doctrine had perhaps emphasized organizational measures, the new one made several conceptual advancements as well. It described the information environment as a synergistic interaction of three dimensions: the physical, with the infrastructures and links of information networks; the informational, representing the actual material being carried by the physical networks; and the cognitive, where the human mind applies meaning to the information, which was described as the most important of the three. It also removes the term "information warfare" from the official lexicon, and while most of the rest of the world still uses information warfare as the most descriptive and commonly understood term for this, the DOD has now officially dropped it. The new Joint Publication 3-13 explicitly links IO to the DOD's efforts to transform itself, and it emphasizes the importance of IO's multinational and coalition elements. The role of STRATCOM as the chief advocate and proponent for IO was also emphasized, and its mission of coordinating IO across geographic areas of responsibility, such as between combatant commands in Europe and Asia and across functional boundaries, was described in greater detail than before. The relationship between strategic communication and IO is also stressed, and it provides a definition, albeit a misleading one for information superiority (JP 3-13 2006).

In addition to these high-level strategic changes in IO policy across the DOD, the U.S. military services have also either published or revised their doctrines for IO in the last few years. The Marine Corps published *Marine Corps Warfighting Publication 3040.4, Marine Air-Ground Task Force Information Operations* in July 2003.

The Army published *Field Manual (FM) 3-13 Information Operations* in November 2003. The Air Force published *Air Force Doctrine Document (AFDD) 2-5 Information Operations* in January 2005. All of these new directives reflected their individual service's perspectives on warfare and IO and viewed IO through the lenses of air, land, or naval warfare. The final policy action with respect to the DOD to be discussed came in late 2006, when the Air Force claimed cyberspace as one of its three core operational environments. While some saw this as nothing more than a turf grab for new missions and resources, the Air Force had stated for more than a decade that it operated across three physical environments: air, outer space, and cyberspace. In December 2005, the U.S. Air Force chief of staff, Gen. Michael T. Moseley, and the secretary of the Air Force, Michael Wynne, issued a new mission statement declaring that cyberspace was a core mission area for the Air Force, and they followed this policy statement in the late summer of 2006 with actions to create a major command for cyberspace operations that will stand alongside both the Air Combat Command and Air Force Space Command (Bennett and Munoz 2006).

Even with the major emphasis by the DOD on the *IO Roadmap*, the most radical changes have not occurred in the traditional realms of IO; instead, they have occurred in the more nebulous regions, such as strategic communications, public diplomacy, international public information, perception management, and psychological operations. A logical place to review the recent IO policy changes in these areas will involve the federal interagency cooperation and coordination efforts. This is because while no National Security Presidential Directive (NSPD) has been released on strategic communications or IO, a significant number of new strategic guidance directives have been published. These include "National Military Strategy," which was published in 2005, and "National Military Strategy on Cyberspace Operations." All of the strategies recommended in these publications require significant coordination across the different federal agencies. Likewise, other key national strategies addressing cyber security, homeland security, and critical infrastructure protection have also been approved, which help to give an overarching framework to IO. While the two policy coordination committees created by NSPD 1 remain in existence, in April 2006 the new Public Diplomacy and Strategic Communication Policy Coordinating Committee was created. It was chaired by the Under Secretary of State for Public Diplomacy and Public Affairs, Karen Hughes (DOS 2006). What is significant about Hughes is her proximity and close working relationship with President George W. Bush.

All of these policy changes emphasize the need for a greater perception management capability within the federal bureaucracy. This idea is crucial because military, political, and economic power is often ineffective in dealing with new kinds of threats to the national security of the United States. The 9/11 attacks were a blow to the American public and its perception of the government, and the fear produced by the terrorist acts can only be defeated by using a comprehensive plan in which information is a key element, or as John Arquilla and David Ronfeldt argued, the concept of networks fighting networks (Armistead 2007). Both Operation Enduring Freedom and Operation Iraqi Freedom represent campaigns fought about perceptions, and the side that will ultimately emerge as the victor is the one that can best shape and influence the minds of not only their adversary, but also the minds of their allies, neutrals, and uncommitted parties. The changes are truly revolutionary and describe a profound shift in the nature of power. Unfortunately, this transformation has not been translated from a strategic concept to tactical actions (Kuusisto and Armistead 2004).

CRITICAL INFRASTRUCTURE PROTECTION

Not all interviewees on this project focused solely on offensive IO policy; indeed, much of their energy and enthusiasm also centered around defensive IO policy. In the last decade, good examples of the development of IO defensive policy can be seen primarily in the operational areas of information assurance and critical infrastructure protection, and in the next section, all of the updates to federal IO policy in these areas will be discussed, as well as their relationship to the desires of the interviewees analyzed.

Critical infrastructure protection is an issue within IO that began with the 1998 issuance of PDD 63 by the Clinton administration. The events of 9/11 affected this area greatly, and the Bush administration followed this initial effort with several policy and organizational changes of its own. Although the *National Strategy to Secure Cyberspace* was issued after the terrorist attacks, the strategy was written and coordinated before that date and reflected the efforts of the Bush administration's then-adviser for infrastructure protection, Richard Clarke, who also had a major hand in the Clinton administration's efforts in this area. In 2003 the Bush administration issued two more strategies, the *National Strategy for the Physical Protection of Critical Infrastructures and Key Assets* and the Homeland Security Presidential Directive (HSPD) 7, "Critical Infrastructure Identification, Prioritization, and Protection." All of these strategy and guidance documents reflected the same basic philosophy of the earlier PDD 63, namely, that the task of conducting information assurance and critical infrastructure protection at the national level was too difficult for unilateral government or business-sector solutions and thus required a partnership between all parties: owners, users, and the national security apparatus. The DOD had recognized the importance of this issue earlier and was one of the principal instigators of a series of national-level studies that began in the early 1990s. Within the Joint Staff, which is responsible for DOD-wide communications, the J-6 directorate has been one of the central players in this area. In early 2006, the J-6 created a new office, the J-6X, and assigned it the responsibility of developing a national military strategy to secure cyberspace, a strategy that was written in the 2001–2002 timeframe. Although this effort was unable to meet its initial timeline of one hundred and twenty days from start to finish, by the end of 2006, the new *National Military Strategy for Cyberspace Operations* had been signed by the Chairman of the Joint Chiefs of Staff.

Likewise, in 2001 under Executive Order 13231, "Critical Infrastructure Protection in the Information Age," the Bush administration redesignated the Committee for National Security Standards as the primary group to provide a forum for the discussion of policy issues; set national policy; and promulgate direction, operational procedures, and guidance for the security of national security systems through the Committee for National Security Standards Issuance System as shown in the Committee for National Security Standards documents 4011–4016. Also from a critical infrastructure protection standpoint, there was an equally prodigious output of directives and memorandum from the Clinton and Bush administrations over an eight-year period, including three executive orders (13010, 13064, and 13231), HSPD 7, and three Government Accountability Office (GAO) reports all issued in close proximity (March, April, and May 2004), to support this area of information assurance. What all of these disparate elements of the business and governmental interests did was to move forward critical infrastructure protection as a vital and useful component of IO. However, because most of the infrastructure portion of critical infrastructure protection is predominantly owned and operated as a function of the commercial sector, progress has been uneven, with some segments, notably banking and finance, advancing more rapidly than others. This disparate focus is especially noted in the three GAO reports that highlight deficiencies in the efforts of the business sector and the federal government.

COMPUTER NETWORK DEFENSE

In addition to critical infrastructure protection, the development of additional policy with regard to computer network defense has been a major component as part of a broader discussion by the DOD on the alignment of IO into offensive and defensive capabilities that match better to their functional organizations. For if international public information (Clinton administration) or strategic communication (Bush administration) are normally considered the "offensive" aspects of this warfare area, then information assurance and its related functions of critical infrastructure protection and computer network defense are more in the defensive realm. In fact, the forerunners of information assurance in the form of information security and computer security have long, distinguished histories within the defense bureaucracy. A good example of this regards a portion of information assurance that centers on computer security assessments as well as the certification and accreditation process. This original methodology for information assurance was known as the *Department of Defense Information Technology Security Certification and Accreditation Process* (DITSCAP 1997), which was in existence for ten years and was replaced in late 2007 by a new certification and accreditation policy titled the *Department of Defense Information Assurance Certification and Accreditation Process* (DIACAP 2007). What this new process does is force the program managers to evaluate their system from a confidentiality, integrity, and availability standpoint on the value of the information protected. To do this, the program managers must determine the confidentiality, robustness, and mission assurance category of their architecture by discussing and analyzing the system with key personnel, such as the user representatives, system administrators, information system security managers, and certification agent. This doctrine was a concerted attempt by the Office of the Secretary of Defense to lay out a new methodology for ensuring the security of its networks and applications by standardizing the process through well-recognized information assurance controls. This is important because this new policy tightens the protection of the government and the DOD by enforcing standards across the enterprise.

There have also been other strategies on computer network defense, such as the *National Strategy to Secure Cyberspace*, which ties into critical infrastructure protection as part of a larger effort to protect America, an implementing component of the *National Strategy for Homeland Security* and complemented by a *National Strategy for the Physical Protection of Critical Infrastructures and Key Assets*. These documents were developed to allow the American public and commercial industries to secure the portions of cyberspace that they own, operate, control, or interact with. Once again, these documents reiterate that IO does not have to be a top-down effort, because power has been shifted to the masses as part of the Information Age, but the responsibility of protecting of America must now be disseminated as well. Citizens of the United States are very accustomed to having the military or armed forces act as their protector against adversaries, but in the Information Age that is not always possible or practical.

DEFENSIVE IO POLICY THAT LED TO THE CREATION OF
THE DEPARTMENT OF HOMELAND SECURITY

As noted above, securing the population is a difficult strategic challenge that requires coordinated and focused effort from our entire society, the federal government, state and local governments, the private sector, and the American people. That is what is different about this current era and what must be accepted in order to truly understand the power inherent in information. The final new policy and organizational initiative from a defensive IO perspective has actually been the creation and development of the DHS. During the fall of 2000 and the spring of 2001, a fourteen-member bipartisan commission headed by former senators Gary Hart (D-CO) and Warren Rudman (R-NH) released a three-part series on the new threats to national security. Named the U.S. Commission on National Security/21st Century, its initial report, "Road Map for National Security: Imperative for Change," attempted to summarize, based upon the changing environment, the new threats to the United States, especially with respect to information (U.S. Commission on National Security/21st Century 2001). These reports proposed radical changes in the structures and baseline processes of the governmental apparatus to ensure that America did not lose its global influence or leadership role.

In an eerie coincidence, the recommendations made by this group provided much of the foundation for the changes that occurred after the attacks of 9/11. While initially scoffed at by academia and the federal bureaucracy, the suggestions of this commission on national security foreshadowed much of the changes that have occurred recently. Equally eerie was a series of comments made by then–CIA director George J. Tenant before the U.S. Senate Select Committee on Intelligence on February 7, 2001. In this testimony, Tenant stated "the threat from terrorism is real, it is immediate, and it is evolving. . . . Terrorists are also becoming more operationally adept and more technically sophisticated . . . for example, as we have increased security around government and military facilities, terrorists are seeking out softer targets that provide opportunities for mass casualties" (U.S. Senate 2001).

U.S. IO Policy: Problems and Successes

Even with all of these official documents and changes in the IO policy and organization within the U.S. government, there have still been a number of issues that have proved difficult to resolve with regard to IO. The problem, as acknowledged by many IO authors and theorists, is that the building of the actual respective steps of the day-to-day tactical operations of IO from the lofty aspirations of IO theory is very difficult. A number of participants in this study alluded to the need for centralized authority and the requisite willpower from the federal authorities to make these dreams of IO theorists come true, yet there were also a significant number of participants in this research project who advocated a bottom-up approach, which could work just as well. For it may indeed be a long time before the U.S. government organizational, personnel, and doctrinal changes catch up to the conceptual power of information, which was lauded nearly a decade ago as the term "IO" first became popular. So in the broadest sense, a disconnect still exists. This can be seen in the initial rush of excitement about information warfare and the revolution in military affairs in the 1995–1996 time frame. While the development of this relatively new concept continued unabated, and a number of exercises were conducted during this period, there was still a gap in the performance of IO. The computer network attack operations conducted during the 1996 and 1997 exercises were particularly effective and drew attention to the fact that the DOD was vulnerable to this type of operation (Pilecki 2002). But as the next two case studies will demonstrate, there is still much work to be done. While some areas of IO have progressed well, there are other areas that have not progressed as satisfactorily as hoped for a variety of reasons.

For example, over the last few years, much has been written on the potential threat posed for the targeting of computer networks and related infrastructures by individuals or groups for terrorist purposes. However, much of this literature has been sensationalist, focusing narrowly on technical computer security issues, and has failed to link the discussion of cyber terrorism with the broader issues relating to either terrorism or policy responses to it (Devost 2003). It is precisely this interdependence between the changing nature of global terrorism and the increasing vulnerability of the critical infrastructures that makes this topic and issue so important. In this next section, the author examines the development and role of critical infrastructure protection within the U.S. government as it relates to IO; he also compares and contrasts its success to other areas, specifically perception management.

U.S. CRITICAL INFRASTRUCTURE PROTECTION POLICIES PRIOR TO 9/11

During the Cold War, U.S. national security policy was focused on minimizing the possibility of strategic nuclear attack by the Soviet Union. There was a general understanding of the nature of the threat posed by the Soviet Union, and most of the international security efforts of the United States (and the West in general) were directed at countering it. But with the collapse of the Soviet Union in 1991, and with it the relatively static bipolar world order, the strategic certainty provided by this structured threat disappeared. A range of diffuse unstructured threats and challenges replaced the specter of global nuclear war. The reality of the new security environment was brought home to the United States with the bombing of the World Trade Center in February 1993. A little over two years later, the scene was replayed when domestic terrorism struck at the nation's heartland on the morning of April 19, 1995, at the Alfred P. Murrah Federal Building in Oklahoma City.

These events raised awareness of the threat posed by terrorism to the United States, but tangible policy outcomes took a little longer to emerge. The first key Clinton administration response to the evolving terrorist threat was to promulgate PDD 39, "U.S. Policy on Counter-Terrorism." This new doctrine articulated a four-point strategy that sought to reduce vulnerability to terrorist acts, deter terrorism, respond to terrorist acts when they occur, and implement measures to deny terrorists access to weapons of mass destruction, while integrating both domestic and international measures to combat terrorism. PDD 39 was novel in that it specifically identified the vulnerability of critical infrastructures and potential terrorist attacks as issues for concern. But in general, this new policy lacked sufficient bureaucratic teeth to achieve meaningful outcomes. What the doctrine did accomplish was to raise the profile of infrastructure vulnerability in the U.S. government, because previous critical infrastructure protection policy had tended to be overshadowed by other elements of national security policy, but this changed in the middle of the Clinton administration (Cordesman and Cordesman 2002, 1–2).

Part of the reason for this rising awareness was the increasing interconnectedness of the Information Age, which has created a range of dependencies and vulnerabilities that was historically unprecedented. Following the 1995 terrorist attacks in

Oklahoma City, the Presidential Commission on Critical Infrastructure Protection (PCCIP) was established by Executive Order 13010. While this group was a natural follow-on to PDD 39, in an informal sense, it also consolidated a range of uncoordinated critical infrastructure protection policy development activities occurring across government (Rattray 2001, 339–340). Likewise, Executive Order 13010 also directed the establishment of an interim Infrastructure Protection Task Force (IPTF) within the Department of Justice, chaired by the Federal Bureau of Investigation (FBI) (Vatis 1998). The purpose of this task force at the FBI was to facilitate coordination of existing critical infrastructure protection efforts under the broad umbrella of the PCCIP. The IPTF was chaired by the FBI so that it could draw upon the resources of the Computer Investigations and Infrastructure Threat Assessment Center (CITAC), which had been set up there in 1996 (Vatis 1998). In essence, the IPTF represented the first clear effort to establish coordinating arrangements across different government agencies and within the private sector for critical infrastructure protection.

In the final report by the PCCIP in 1997, this group produced a document titled "Critical Foundations," whose key finding noted that while there was no immediate overwhelming threat to the critical infrastructures, there was a need for action, particularly with respect to the protection of the national information infrastructure. The report also recommended a national critical infrastructure protection plan, with clarification of legal and regulatory issues that might arise out of such a plan and a greater overall level of public-private cooperation for critical infrastructure protection (PCCIP 1997). To follow through on these findings, between late 1997 and early 1998, PCCIP underwent an interagency review to determine the Clinton administration's overall response to this policy initiative (Moteff 2003, 4). Even as that was underway, concrete outcomes were already beginning to emerge by February 1998 as the interim IPTF was amalgamated with CITAC and made permanent within the FBI under a new name—the National Infrastructure Protection Center (Vatis 1998).

The recommendations of the PCCIP were also given practical expression on May 22, 1998, with the release of two policy documents: PDD 62, "Counter Terrorism," and PDD 63, "Critical Infrastructure Protection." These two documents were the culmination of the Clinton administration's efforts at policy development for counterterrorism and critical infrastructure protection. This new directive by the Clinton administration also provided a more defined structure for counterterrorism operations and represented a focused effort to weave the core competencies of several agencies into a comprehensive program. Also in common with PDD 39, PDD 62 sought to integrate the domestic and international elements of U.S. counterterrorism policy into a coherent, whole structure.

PDD 63 was also the document that implemented the recommendations of the PCCIP report, as interpreted through the prism of that interagency review panel. Identifying twelve sectors of critical infrastructure protection that needed additional support, this directive appointed government lead agencies for each of these sectors and established coordination mechanisms for the implementation of these measures across the public and private sectors. In particular, PDD 63 vested principal responsibility for aligning these activities in the Office of the National Coordinator, which had been set up under PDD 62. PDD 63 also established the high-level National Infrastructure Assurance Council to advise the president on enhancing the public and private partnership for critical infrastructure protection. In addition, this directive called for a National Infrastructure Assurance Plan, which would mesh together individual sector plans into a national framework. Finally, this document also authorized increased resources for the National Information Protection Center and approved the establishment of sector information sharing and analysis centers to act as partners to the NIPC.

There were also additional updates in the last year of the Clinton administration that included minor changes to critical infrastructure protection policies. Version 1.0 of the "National Plan for Information Systems Protection" was released in January 2000, as a direct result of the call in PDD 63 for a National Infrastructure Assurance Plan (Executive Office of the President 2000; Moteff 2003, 19). It is interesting that given the priority reflected in cyber security issues by the PCCIP, the National Plan primarily addressed the national infrastructure protection rather than critical infrastructure protection as a whole (Executive Office of the President 2000). As noted in PDD 63, critical infrastructure protection cannot be limited to just the federal infrastructure because the public sector cannot be separated from the private sector in today's information environment. Other changes also occurred in the waning days of the Clinton administration. In June 2000, the Terrorism Preparedness Act established the Office of Terrorism Preparedness within the Executive Office of the President. Its role was to coordinate counterterrorism training and response programs across federal agencies and departments. Like the Office of the National Coordinator established by PDD 62, the Office of Terrorism Preparedness was not granted budgetary authority and often had to rely on persuasion rather than a formal chain of command to achieve its objectives.

When the second Bush administration came to power in early 2001, there was some consolidation of existing critical infrastructure protection arrangements. The collection of senior critical infrastructure protection groups was consolidated into one Counter-Terrorism and National Preparedness Policy Coordination Committee that reported to the NSC (Moteff 2003, 8). And while some debate occurred on future directions for counterterrorism and critical infrastructure protection policy, these bore no fruit prior to the 9/11 terrorist attacks (Moteff 2003, 8). So in practice, during the first nine months of the second Bush administration, the bulk of the counterterrorism and critical infrastructure protection arrangements in place in the United States were largely a legacy of the Clinton administration.

Thus to summarize, in the decade prior to the 9/11 attacks, when the international aspect of the terrorist threat to the United States was becoming more evident, significant policy updates were being promulgated by the White House. These terrorist incidents that demonstrated the international character of the terrorist threat included the February 1993 World Trade Center bombing, the June 1996 attack on the Khobar Towers complex in Saudi Arabia, the plans to attack U.S. airliners in Southeast Asia in 1995, the attacks on U.S. embassies in Kenya and Tanzania, and the attack on the USS *Cole* in October 2000. In response to all of these incidents, PDD 39, PDD 62, and PDD 63 were all incorporated as measures to combat terrorism abroad and protect critical infrastructure domestically. But while the international dimension of the evolving terrorist threat was acknowledged directly in policy, in actuality it was largely overshadowed by the domestic aspects of U.S. policies on counterterrorism and critical infrastructure protection, which were implemented during this period.

PDD-68, "INTERNATIONAL PUBLIC INFORMATION"

It was also during this time frame of critical infrastructure protection development that a major effort by the U.S. government to improve its perception management capability was also begun. Not listed in the original Joint Publication 3-13 policy, perception management is generally considered to be comprised of a number of subelements, including public affairs, influence campaigns, public diplomacy, psychological operations, deception, and covert action. In reality, perception management is simply the ability to shape an image or conduct an influence campaign. Defined by the DOD below, perception management is also seen as a key focus of change within the U.S. government.

> Actions to convey and/or deny selected information and indicators to foreign audiences to influence their emotions, motives, and objective reasoning; and to intelligence systems and leaders at all levels to influence official estimates, ultimately resulting in foreign behavior and official actions favorable to the originators objectives (Joint Publication 1-02 1998, 340).

In addition to the publication of the seminal doctrine of Joint Publication 3-13, the White House and the DOD have also realized that they needed better coordination with regard to IO, since these influence campaigns are often conducted long before the traditional beginning of active hostilities (Metzl 2003). This interaction between federal agencies within the executive branch also brought about a renewed emphasis on developing the correct IO organizational structure. As alluded to earlier in this chapter, the DOS was engaged in a major organizational shift, as the USIA component was brought within the greater cabinet agency. The actual legislation that amended the structure of the DOS is known as H.R. 1757, "Foreign Affairs Reform and Restructuring Act of 1998." Divided into three parts, it is in Division A, Title III-V, where the abolition of the different DOS functions are discussed in detail (U.S. Department of State 1998). What is very interesting is that the actual language of the bill states that its purpose is to strengthen and coordinate U.S. foreign policy by giving the secretary of state a leading role in the formulation and articulation of foreign policy through the consolidation and reinvigoration of foreign affairs functions (U.S. Department of State 1998). To do this, the writers of this bill proposed the elimination of the U.S. Arms Control and Disarmament Agency (ACDA), the USIA, and the U.S. International Development Cooperation Agency (IDCA). By definition, the mission of the DOS is to advance and protect the worldwide interests of the United States (Armistead 2002). However, the USIA was designed to understand, inform, and influence foreign publics as a means of promoting U.S. national interests and dialogue between Americans and their institutions and counterparts abroad with its 7,000 employees (Armistead 2002). The IDCA and ACDA were smaller agencies with very specialized missions, but under this proposal, all of the functions, personnel, and funding from these organizations would be transferred to the DOS to increase the power of the cabinet-level agency.

The DOD and DOS were not alone in the organizational changes with respect to the power of information and perception management. In late 1997 and throughout 1998, the NSC, under the leadership of Richard Clarke, began to develop the framework for what eventually became PDD 68, "International Public Information," policy (Metzl 2003). Originally, not all executive-level organizations agreed on the need for an information policy. Not only did they need to be convinced of its importance but also about the timeliness of this issue (Metzl 2003). To do this, the NSC integrated this new information concept into the larger reorganization effort of the DOS. In addition, DOD officials were also meeting in November 1997 to build a sub-group to support the larger construct of PDD 56, "Managing Complex Contingencies" (Dorflein 2000). This earlier policy document had been signed as a tool to help the interagency process cope with complex contingencies as mentioned earlier, and its main output was the development of an executive committee that would meet and help make executive decisions during a crisis. The problem, as laid out by NSC director Richard Clarke in his "Terms of Reference," was that if one waits until a crisis has occurred to get together and form a committee, then one cannot use the power of information to help shape the environment (Metzl 2003). Instead, at this November 25, 1997, meeting, Clarke suggested that there was a need for the group to develop a process to build a construct that would allow the United States to plan much earlier for an information campaign. Thus the primary task of this interagency group was to study the issue of how the U.S. government used information over the next six months and conduct an assessment of U.S. and multilateral agencies for planning, coordinating, and conducting perception management activities within the context of the PDD 56 construct (Metzl 2003).

What is especially interesting is that the decision to combine public diplomacy and public affairs under the mantle of international public information is totally in opposition to the conclusion that the Truman administration came up with nearly fifty years earlier. In 1948 DOS officials dealing with these same two issue areas thought it was too difficult to coordinate under one office. They split, and the USIA was formed (Armistead 2002). In fact, Congress was so concerned about the possible propagandizing of the American public that they passed the Information and Educational Exchange Act of 1948, which legislated that the DOS could only conduct public diplomacy abroad and only to foreign nationals. Technology has changed dramatically over the last five decades, and the ability to segregate or separate access to information is much more difficult today. For example, how is it possible to ensure that only a foreign audience views an Internet-based website, especially given the fact that video and audio-streaming technology, radio, and television broadcasts can now be sent around the world? Are the changes to information and perception management affecting the nature of public diplomacy? It is these types of questions and many others that had to be answered by this interagency working group as they struggled to find consensus on their new policy.

However, change does not come from policy alone. As most bureaucrats understand, the real power of an organizational change, especially a large one such as at the DOS, often only results from funding and personnel moves (Kovach 2004). Thus it was not until August 2000, more than sixteen months after the original signing of PDD 68, that the first uniformed military officer was stationed at the DOS; it was only at that time that true progress began to occur in moving forward on this initiative (Ward 2001). Former DOS officials, including Jamie Metzl, Peter Kovach, and Joe Johnson, had all done an incredible job of keeping the flame and spirit of international public information alive, but their job was not to function as planners. Therefore, what was truly needed to make this program work was an action officer and staff who could be assigned to run a program. As one of the participants stated, the biggest problem with international public information early on was that there were no operators—that is, no one or no group to operationalize the process—and until they were brought onboard, little overall progress was made (Ward 2001).

IO Organizational Changes in the U.S. government

The changes, or the lack of, at the DOS mentioned in the section above are symptomatic of an overall trend within the federal government toward IO. There have not only been major changes to IO in the form of public policy by the federal bureaucracy; but organizationally, the landscape of IO has shifted dramatically as well. The term "suburbanization" is often used to describe the changing role of IO from an organizational perspective. Ten years ago, with the huge emphasis on the revolution in military affairs and the introduction of information warfare, grand themes and terrible scenarios were described in great detail to the public and Congress alike. These scenarios included Electronic Pearl Harbor, cyber war, and other similar threats that provided a degree of hyped emphasis, which while helping to introduce the vulnerabilities associated with IO, often distracted from the overall goal. This was because these sensationalistic briefs tended to bring about hysteria, which had the unfortunate effect of desensitizing personnel to the real dangers inherent in IO that often tend to be more mundane and technologically complex. For example, early descriptions of cyber attacks often foretold of massive panic as hackers brought down the power grids in the United States. However, when this actually happened on August 14, 2003, in the northeast portion of the United States as a result of a fault in a power plant, panic did not ensue. Instead, millions of people were relieved that it was only a technical hitch and not a terrorist attack. What followed this Electronic Pearl Harbor was not pandemonium; instead, with a bemused attitude and predictable New York spirit, there was a refreshing demonstration of peoples' resilience as they made the long walk home on a hot, powerless day. It is this movement from the sensationalized attitude surrounding IO to a more operational, or suburbanized, effort that best reflects the overall theme of this section. Federal agencies can no longer develop IO solutions alone or in a vacuum, and the changes to IO policy and organizations in the United States tend to become noticeably less profound but more detailed, with more depth and substance, as time passes. What has changed specifically is the awareness that when integrated planning is conducted, its results can synchronize the efforts of many different commands, services, and agencies, so that the value-added benefits of an information campaign quickly become apparent. In addition, because information efforts are often conducted long before the traditional beginning of active hostilities, the need for the White House and the DOD to coordinate between themselves and other government agencies and departments has brought about a renewed emphasis on the information organizational architecture.

U.S. CRITICAL INFRASTRUCTURE PROTECTION POLICIES AFTER 9/11

The terrorist attacks of 9/11 led to fundamental changes to the U.S. government's approach to critical infrastructure protection issues. On October 8, 2001, Executive Order 13228 established the Office of Homeland Security, to be headed by the Adviser to the President for Homeland Security, Tom Ridge, the former governor of New Jersey. The purpose of the Office of Homeland Security was to develop and coordinate a national strategy to protect the United States against the new threats posed by global terrorism. This directive also established a high-level Homeland Security Council, which was responsible for advising the president on all aspects of homeland security (Executive Order 13228 2001). The following day, appointments were made for the National Director for Combating Terrorism, Gen. Wayne Downing, and the Special Adviser to the President for Cyberspace Security, Richard Clarke, via Executive Order 13231. What is significant about these appointments is that Downing had previously been the Commander in Chief of the U.S. Special Operations Command (USSOCUM), so his appointment reflected a greater prominence for the international and overtly military dimension of U.S. counterterrorism policy. In addition, this directive also created the President's Critical Infrastructure Protection Board (PCIPB), whose duty was to recommend policies and strategies for the protection of critical information systems. The same executive order also established the high-level National Infrastructure Advisory Council (NIAC) to provide advice to the president on these key issues (Moteff 2003, 10).

These efforts were not the end of new policy development with regard to critical infrastructure protection in the aftermath of 9/11. In July 2002, the Office of Homeland Security released the *National Strategy for Homeland Security*, the purpose of which was to integrate all government efforts for the protection of the nation against terrorist attacks of all kinds (Moteff 2003, 11). In effect, the strategy updated the measures enacted under PDD 63 in light of the post–9/11 environment. This new strategy did not create any additional organizations but assumed that a DHS would be established in the near future (Moteff 2003). This document was updated in September 2002, when the PCIPB released the draft *National Strategy to Secure Cyberspace*. In effect, this document was the proposed successor to the Clinton administration's *National Plan for Information Systems Protection*. While the issue of the draft plan was welcomed, concerns were expressed that it lacked the regulatory teeth to prompt action by the private sector, which goes back to some of the original faults embedded in PDD 63, namely, there must be a tight coordination between the public and private sectors.

The most obvious consequence of the revised U.S. approach to critical infrastructure protection in the aftermath of 9/11 occurred in November 2002 with the creation of the DHS (Moteff 2003, 11). This new agency consolidated the bulk of U.S. federal government agencies dealing with homeland security, which consisted of over 170,000 employees, into one department headed by a cabinet-level official (Moteff 2003). Representing the most fundamental change to U.S. national security arrangements since their inception in 1947, the DHS is comprised of five directorates:

1. Management, Science, and Technology
2. Information Analysis and Infrastructure Protection
3. Border
4. Transportation Security
5. Emergency Response and Preparedness (Department of Homeland Security Organization 2003)

What is very interesting and significant is that the DHS closely resembled some of the measures that had been proposed by the U.S. Commission on National Security/21st Century (Moteff 2003, 8, 9). But it was only after the events of 9/11 that the political imperative for significant organizational change in the field of critical infrastructure protection emerged. Further action with regard to this IO warfare area also continued within the Bush administration in 2003 when three more policy documents were released:

1. *National Strategy to Secure Cyberspace*
2. *National Strategy for the Physical Protection of Critical Infrastructures and Key Assets*
3. *National Strategy for Combating Terrorism*

At the same time, the release of Executive Order 13286 abolished the PCIPB and the position of Special Adviser on Cyberspace Security (Moteff 2003, 10). The NIAC was retained, but it now reported to the president via the DHS. Combined with the departure of key staff associated with cyber security issues, these measures raised concerns that cyber security issues were being marginalized in the new arrangements (Moteff 2003, 23, 24).

In summary, what this section lays out is the evolution of critical infrastructure protection within the U.S. government. Conducted in fits and starts, it is often only with tremendous political pressure that many of the changes recommended by these different blue ribbon committees and groups have been adopted. However, there is still more to do, because so much of critical infrastructure protection is tied to the partnership between the public and private sectors. No matter what is promulgated on the federal side, until the corporate executives are convinced of the return on investment from these initiatives, the true potential of these directives will not be realized. For that is a key point missing from some of these publications and emphasized by the research interviews; namely, that critical infrastructure protection cannot be mandated to the business world. Instead, an education campaign must be conducted to show why these efforts are justified. To date, the author does not believe, nor does the literature show, that this training has occurred.

THE EFFECTS OF 9/11 ON IO ORGANIZATIONS

The events of 9/11 were a tremendous wake-up call for the Bush administration in regard to how it conducted IO at the executive level. In the days immediately after these attacks, the DOS was looking to the executive branch and the NSC for guidance on building an organization to support a strategic information campaign. Unfortunately, leadership was slow in forming. In the period after the terrorists' strikes, there was a significant amount of confusion within the government, and this paralysis carried over to the conduct of IO. For the first five to six weeks at the NSC, there was an absence of both knowledgeable and experienced people to deal with strategic influence campaigns, as well as the normal intraorganizational discontent and turf battles (Jones 2003). At that time, the Clinton-era NSC document, PDD 68, "International Public Information," had been effectively muted, so there was no office at the NSC to conduct a strategic perception management effort. During major portions of this crucial period, the Joint Staff ended up simply contracting out their perception management campaign to the Rendon Group, a civilian company that specializes in strategic communications, under a contract with the DOD (Jones 2004). Gradually, as the campaign on terrorism continued throughout the fall of 2001, a number of influence plans and strategies were developed to create a working operational group, yet the NSPD still remained in a holding pattern within the interagency process.

In November 2001, in accordance with NSPD 8, which established the Office of Combating Terrorism and outlined General Downing's roles as Deputy Assistant to the President and U.S. National Director and Deputy National Security Adviser for Combating Terrorism, a new position of Senior Director for Strategic Communications and Information was created which helped to bring a level of competence to the staff (Jones 2003). Likewise, during the immediate aftermath of the terrorist attacks, Alistair Campbell, the Communications Director for British prime minister, Tony Blair, had suggested forming a series of Coalition Information Centers to concentrate on getting the pro-American message to the world media. Eventually, three of these centers were set up: one in Washington, D.C., one in London, and one in Islamabad. The facility in Pakistan actually occupied an old USIA building. All together, these groups perform admirably, focusing on public affairs and public diplomacy. However, some critics argued that these organizations concentrated on U.S. domestic partisan politics instead of focusing on the set of global audiences now accessible via a 24/7 news environment (Armistead 2003). Other critics have argued that these Coalition Information Centers generally did well at informing domestic and foreign press within their time cycles during the early phases of Operation Enduring Freedom, and they also eventually utilized a U.S. government spokesman who could speak Arabic and thus appear live on the Al Jazeera television station. Looking back, it is uncertain whether this really was a success story, because the question remains of why it took so long for Ambassador Christopher Ross to appear on Al Jazeera. This may have been because the White House was slow to see the need for U.S. presence on Al Jazeera until external pressure became such that Colin Powell and Condoleezza Rice were forced to appear on this Arabic TV station using translators. In fact, Al Jazeera constantly invited them for interviews early on, but these invitations were rebuffed, and Al Jazeera was actually blacklisted from early White House press conferences. Eventually, the response was changed, but this audience was crucial, a fact that should have been recognized much earlier (Rendon 2003). Foreign media always needs to be addressed in this Global War on Terror (GWOT). That it took so long to make key U.S. government personnel available to these media sites may be an indication that the U.S. government did not at first understand the true nature of this new battlespace.

Yet all was not bleak. Before Karen Hughes left the Bush administration in its first term, she formed the Office of Global Communications, ostensibly to force the public diplomacy community resident within the DOS and in the field to do a better job of explaining overall U.S. policies (Armistead 2003). Created out of frustration with the perceived lack of effort at Foggy Bottom, this office coordinated with the interagency Global Communication Strategy Council. An evolutionary process and a follow-up to the Coalition Information Center, this White House staff also coordinated with the NSC in a quid pro quo relationship. Yet the departure of Hughes and General Downing from the Bush administration probably spelled the ultimate demise of the Office of Global Communications and further White House strategic communication efforts (Alter 2002, 49). However, there are those who don't believe this was Office of Global Communications mission at all. They believe its real task was to be the influence arm of the White House and get President Bush's message out as an element of his reelection campaign for 2004. While this would be a normal and understandable objective of any White House-based communications effort, suspicions remain that the then-director of White House communications, Karen Hughes acted quickly in early 2002 to put the new Strategic Communications Policy Coordinating Committee on hold because of fears that it would interfere with this mission (Jones 2003). The fact that shortly after the election of November 2004, this Office of Global Communication quickly and quietly ceased operations could be a support for this interpretation.

This emphasis on the domestic audience can also have negative effects in other ways, too. First, there appeared to be a lack of understanding about what words or phrases mean to other audiences. For example, some may be instantly hostile to an Islamic audience, while others may have an impact poorly understood by Western-ers. "Axis of Evil," "Infinite Justice," and "Crusade" are great examples of the Bush administration's public diplomacy missteps. In addition, the White House did not collaborate well with DOS specialists who understood the implications of such phrases, and their misuse of these actions and words, which seriously hurt the Bush administration in its early phases of the GWOT. Some quip that a serious review of Samuel Huntington's *Clash of Civilizations* is not out of the question. Likewise, the use of common Islamic terms to label our adversaries may have a negative and unintended consequence. For example, including suicide bombers and terrorists under the label "jihadists" may have actually been seen as legitimizing them and their actions. Labels and terms are used in many cases because they are easy and in the common lexicon, yet it is often not understood how they appear and what they mean in other cultural contexts. In a "war of ideas," words can be as lethal as real ammunition (Armistead 2007, 158).

The IO organizational changes at the interagency level only became more con-voluted as the GWOT continued (Foer 2002). The J-3 Director of Operations on the Joint Staff formed the Information Operations Task Force, led by the J-39, to be responsible for IO. However, that group was more technically oriented, so there was still a role for the DOS in the diplomatic arena (Pilecki 2002). A Strategic Informa-tion Core Group was also formed within the interagency structure, but the general consensus was that not much was accomplished with this organization because it was never empowered or recognized by the major departments to possess the abil-ity to get things done. In this atmosphere of Operation Enduring Freedom and the ongoing war in Afghanistan, the DOD established the OSI in November 2001 in an effort to coordinate its strategic perception management campaign. There was a perceived leadership void that led to the Assistant Secretary of Defense for Special Operations/Low Intensity Conflict being put in the lead. The OSI organization was comprised mostly of personnel with psychological operations and civil affairs backgrounds. Its mission was to respond to and negate hostile propaganda using mostly human factors and a little technology (Timmes 2002). It appeared to be placed to work well because it had financial resources and was a DOD organization, yet it quickly ran afoul of two critical interagency IO organizations (Rötzer 2002c). This is because the OSI group had been placed at the DOD, not at the DOS's Bureau for International Information Programs. This happened because some believed that its more operational tasks would be more easily accomplished from within the DOD. By doing this, the DOD gave the ultimate rejection to PDD 68, which may have stemmed from the overall belief that the strategic perception management campaign had been wrongly placed by the Clinton administration, and that an office should have gone to the DOD or NSC instead.

At a meeting on February 16, 2002, Secretary of Defense Donald Rumsfeld approved the office. However, the senior DOD Public Affairs official, Victoria Clarke, did not concur, and her opposition manifested itself almost instantly. On February 19, 2002, the first article critical of the new organization appeared in the *New York Times*. It was released while both Rumsfeld and Clarke were in Salt Lake City, Utah, at the Winter Olympics. It was reported that Rumsfeld was livid but could not do much owing to the political concerns created by the allegations that the OSI would lie to the media to conduct disinformation campaigns. As satirically reported by Mark Rodriguez in the *Washington Post* electronic journal *Insight*, the demise of this DOD office was a political turf–battle. Clarke was leading her own disinformation campaign to retain control of all public affairs efforts, exactly the charge she made to the press about OSI, which was later investigated and proven unfounded (Ricks 2002). Politically embarrassing to Rumsfeld and Bush, it was very comical to watch the government officials deny the need for an office in the United States to conduct strategic perception management campaigns. Every nation participates in these activities, but almost all deny their existence. Even foreign news agencies put a satirical touch on their reporting as they watched the American officials attempt to explain away the obvious (Woodward and Balz 2002; Rötzer 2002a; Creveld 2002).

All of these organizational shifts with regard to strategic communications allude to a question that has arisen over the last ten years, namely, where should a strategic perception management campaign office be located? PDD 68 put the international public information activities at the DOS in 1998 where it foundered for two years owing to a lack of budgetary authority, manning, and empowerment. In addition, the international public information group was also hampered by the interagency process. While the draft NSPD on strategic communications has repeatedly recommended the need to embed the strategic perception management capability into an office in the NSC, in 2001 the Defense Science Board's (DSB) report, "Managed Information Dissemination," reiterated the desire to keep the authority at the DOS (Gregory 2003). This argument for keeping the Policy Coordinating Committee at the NSC was centered on the desire to keep this organization in a steady state. The NSC is, by definition, the single organization within the U.S. government responsible for turning interagency positions into recommendations to the President. It looks at international affairs and foreign audiences in an operational manner, a viewpoint that was missing from the international public information way of doing business. So there is strong logic behind this argument as well. The counter-prevailing suggestion for putting the Policy Coordinating Committee in the DOS was led by David Abshire, who believed that a Tom Ridge-like figure was needed to drive the program (Fulton 2003). However, there is also a concern that any strategic communications effort led by the DOS will be focused more on public diplomacy and public affairs than on strategic influence issues.

This effort was eventually overcome by events. With the initial departure of Karen Hughes from the White House in 2002, most of these activities lost their momentum. After all, it was Hughes who made the Coalition Information Centers happen during the early stages of Operation Enduring Freedom. She understood how effective public diplomacy could be in the GWOT. The Coalition Information Centers were so successful during the fall of 2001 because of the president's influence, and also because there were no constraints. In effect, they didn't have to filter information through a number of layers of bureaucracy as is normally the case because Congress is very concerned with the U.S. Information and Educational Exchange Act of 1948 (i.e., the Smith-Mundt Act) (Gregory 2003).

As the events surrounding the OSI debacle of early 2002 indicated, the widespread concern toward activities of the DOD and the DOS may have not been the case when it comes to the White House. With the creation of the Office of Global Communications and its assigned mission of explaining the U.S. policies, the White House felt a great need during Operation Enduring Freedom to expand its frame of reference. For example, it wanted the ability to influence those Islamic nations and populations that reject out of hand any information coming from western sources. This theme was emphasized by David Hoffman of Internews Network in his *Foreign Affairs* article, "Beyond Public Diplomacy," when he asked the quintessential question, "How can a man in a cave out-communicate the world's leading communications society?" This question reinforces the need for more concerted strategic communication efforts by the U.S. government (Hoffman 2002). Therefore, the DOS still needs to enlist moderate Arabic nations to help in this project, but this desire runs into the roadblock of how current American efforts in the Israeli-Palestinian conflict are seen across the Islamic world—and exploited by Islamic radicals, sometimes via overt disinformation—as clear evidence of a United States-Zionist alliance. The conflict in southern Lebanon in the summer of 2006 merely added fuel to this fire. Often, the U.S. government does not necessarily see the connection between the Palestinian conflict and events in Iraq, but the entire Arabic world does instantly. So now the White House is even losing out on trying to get the moderates to push our message. Plus, the debacle concerning the OSI in February 2002 also stalled any of the subsequent Bush administration's attempts to develop a strategic communication effort, and essentially this controversy put the NSC's Strategic Communication Policy Coordinating Committee on hold until the creation in April 2006 of the new Public Diplomacy/Strategic Communication Policy Coordinating Committee, chaired by Karen Hughes (Armistead 2007).

Thus the mission and structure of the new Policy Coordinating Committee constitutes an attempt by the Bush administration to develop a long-term capability to conduct public diplomacy and strategic communication. There is still no overarching U.S. government strategy for strategic communications, despite the fact that the White House has had a counterterrorism information strategy since December 2001. There can be little doubt that the proposed strategy was an attempt to answer this long-sought government-wide effort. The irony is that it was only in 1999 when the USIA was dismantled and its functions shifted under the greater umbrella of the DOS. In fact, as mentioned previously, Representative Henry Hyde (R-NY) proposed numerous times the reconstitution of that agency in his legislation to bring back capabilities that had so recently been diminished, for much of this legislative proposal mirrors efforts by the DSB's "Managed Information Dissemination" working group. While the DOS did not agree with this concept, the new structure suggested by the Karen Hughes-chaired Policy Coordinating Committee may go even beyond what existed previously in terms of a strong, centrally influenced communication program. Therefore, the demise of the USIA may have contributed to the failing of PDD 68, and thus the need for a new structure and capability to conduct global influence, more than any other action to date (Ward 2001).

In the end, it is not a new organization that will drive a strategic communications effort; instead, there needs to be a shift in the mindset of the White House and the NSC. The need to push senior officials to conduct briefings to match Middle Eastern news cycles, or to ensure U.S. Arabic speakers are available on Al Jazeera, is becoming much more accepted and understood methods of doing business. These ideas are now conventional wisdom as the value of strategic communications rose within the Bush administration. To be effective, the government cannot just think in news cycles (24/7 around the world); instead, it must think in decades. One way to do this is to expand exchange programs, such as the Fulbright Scholar Program, so that the U.S. government can be much more effective in a strategic management campaign. This latter example could be an example of one of Karen Hughes's "Four E's" of public diplomacy: engage, exchange, educate, and empower. In effect, there needs to be an issues agenda versus a values agenda. We need to take a short- and long-term approach to these problems, but it must also be led from the top down, with full White House and NSC leadership to ensure full interagency participation (Jones 2004). The second Bush administration repeatedly tried to "talk the talk" of public diplomacy and strategic communications. Quotes and sound bites referring to the need to do these tasks better abounded from all levels, but what is really needed now is to "walk to match the talk." The government needs to provide real evidence of resources, organizations, people, and operations that enable an effective long-term strategic communications campaign. It is only then that a true strategic perception management campaign will succeed and the power of IO be realized by the United States.

Summary

What all of these reports emphasize is the need for a much greater capability with regard to perception management and strategic communications within the U.S. government. The mere fact that these publications continued to be released means that the progress envisioned by these various advocates of IO has not materialized. In examining these studies and recommendations of the official U.S. government IO efforts with respect to the GWOT, it is interesting to compare these reports to a series of articles compiled by the *Washington Quarterly* and edited by Alexander Lennon titled *The Battle for Hearts and Minds: Using Soft Power to Undermine Terrorist Networks*. Published in 2003, these articles attempt to show how useful information can be to the United States for campaigns such as Operation Enduring Freedom and Operation Iraqi Freedom. While it will be interesting to see if any of the recommendations of either the semiofficial or commercial publication make it into the next update of Joint Publication 3-13 or other official IO policy, it is fascinating that a number of these articles in Lennon's book advocate the potential of perception management for future operations, and that its proper conduct will be key to success in the future. It was also noted by Lennon that in the greater umbrella of IO, it is the area of perception management that is the most rife with confusion and misinterpretation, because there is such a fine line between psychological operations, public affairs, influence campaigns, public diplomacy, international public information, strategic communications, and propaganda (Lennon 2003).

From a different perspective with regard to perception management in the U.S. government, in Nancy Snow's two books, *Propaganda, Inc.* and *Information War*, she argues that the U.S. government has too much power and uses that power to control society by limiting dissenting opinions and free speech, especially in the Bush administration after the events of 9/11 (Snow 1998, 2003). This opinion is not widely shared by the participants of this dissertation, but all views are valid and should be taken into consideration as part of the methodology of this research. In addition, while the author attempted to select a diverse group of interviewees, that was not always possible in some cases because a high level of IO knowledge was a key factor. So it is very interesting to get different opinions on the use of IO within the U.S. government from authors such as Nancy Snow.

Thus, to summarize this section, of the two areas of IO policy of the U.S. government that were selected to analyze in detail—computer network operations and perception management—it has been the former that has been more successful in its implementation over the last decade.

CHAPTER 5

IO APPLICABILITY TO THEORY AND PRACTICE

In this section, a broad comparison will be conducted to evaluate the differences between rhetoric and reality, especially in the evaluation of the employment of IO across the federal government. The hypothesis in chapter 1, "Understanding the Problem," stated that in the United States, a significant gap exists in regard to the conduct of IO. While this warfare area is a relatively newly defined activity, it has the potential to transform the traditional uses of power and revolutionize the manner in which war, diplomacy, business, and a number of other areas are conducted. Yet all too often, hyperbole and unrealistic desires hamper actual progress of this concept. The analysis of this gap between the proposed capabilities and the actual conduct of IO missions operations is the main thrust of this book. Specifically, as part of this research, a number of examples surfaced during the interviews to validate the research hypothesis as well as provide new information regarding the usefulness of IO with respect to the U.S. government.

Why does IO matter?

One of the key goals of this research is to evaluate the gap between stated goals and actual operations of IO across the U.S. government by using a qualitative interpretative approach via a systems process, specifically soft systems methodology. A total of fifty-four interviews were conducted over a five-year period with forty participants to produce two very divergent conceptual models. These can be viewed as polar opposites of one another, with one school of participants advocating a top-down approach as the best method to conduct IO, and other interviewees instead declaring that the only way to make any progress in this particular area was via a bottom-up approach. This latter idea became a key point of this research, primarily because a significant number of participants believed that they were simply echoing a more realistic view of what makes the power of information so unique. For unlike the traditional loci of power (military, diplomacy, and economic), all of whose instruments the government normally controls, this is not the case with IO. Specifically, the power of information lies with the individual, as do the controls and tools. This is an extremely radical and a salient feature of IO—the government can no longer control information; instead, this element of power has been disseminated down to the masses. This inability to control this element of power, or to even understand that the government is no longer in control of information, is perhaps the most important point in this whole research. It was repeatedly shown in the book interviews that the enlightened government officials who understood this concept—that the flow of information could be influenced but not not dominated—were the organizations in the federal bureaucracy that fared relatively well in this new environment. It was also demonstrated that those agencies and staff that refused to acknowledge the seismic shift that had occurred with regard to power and information were ultimately the ones that repeatedly were unable to compete in this rapidly advancing field.

Taken together, all of these data points have been combined into four key themes that are part of an overall analysis of the major deficiencies that succinctly articulate not only why the aforementioned gap in the performance of IO exists, but also what approaches could be useful in helping formulate a way ahead for more successful efforts in the future.

1. Why is there no overall strategic theory in the United States for IO?
2. Is IO really the best term to describe these activities?
3. Why is the top-down approach to IO not working in the U.S. government?
4. Why is there no rhyme or reason to the IO training and education curricula?

These four critical areas are the main focus of the final analysis of research in this book. For example, the first question was derived from the deficiencies cited by the book interviewees who were concerned about the lack of an overarching theoretical construct for IO. Some participants posited that if the Information Age is truly as radical as many suggest, there should be a more vigorous academic debate with a number of theories vying for ascendancy in this new era. Yet to date, no single comprehensive theory on IO has fallen into general acceptance across the U.S. government. For while strategic military IO policy and doctrine has been promulgated by the DOD, no corresponding similar policy is being developed across the other interagency organizations. The second question arose from the problem that because the actual definition of IO is virtually all inclusive, it is still considered to be vague and barely understandable, and some of the research participants believed that harm is being imparted to IO as a concept by the broader academic, military, and diplomatic community. Information is and always has been a vague term, but in this new era, it possesses a capability that is now considered crucial to the success of U.S. national security; therefore, the proper definition and taxonomy are crucial to success. Another question came from the fact that in most cases, the actual conduct of, or approach to, IO activities and campaigns are normally performed at a more tactical level or in a bottom-up fashion versus a centralized manner. However, there are still many questions about the preferred method by which the U.S. government can most successfully utilize this element of power. Many of the interviewees noted this dichotomy in the fact that because IO crosses so many boundaries within the interagency processes, it is often very difficult to quantify exactly what constitutes an information campaign. As a result, success is often measured in different ways. Finally, the last question arose from the sheer number and diverse quality of IO training and education efforts across the federal bureaucracy, which has led to much inefficiency that corresponds to the inability of the United States to maintain a professional corps of personnel. Specifically, there is no coordination between these different schools of thought. There are no training standards, certifications, or linking mechanisms to show a synergy of effort. This lack of synchronization across the different agencies, commands, and organizations is severely hampering the overall ability of these groups to conduct IO. Thus to summarize, it is these four concepts that continue to highlight the gap between higher-level strategy and operational reality as discussed in the hypothesis. The reasons for this gap have been examined in previous sections of this book, and specific factors will be noted as to why the federal bureaucracy is unwilling or unable to make the transformational changes that are needed to best utilize information as an element of power. It is hoped that these conclusions and recommendations may be useful for future IO planners and senior-level decision makers in the U.S. government.

Why is there no overall strategic theory in the United States for IO?

The problem is that without a strategic theory or academic model to serve as a basis to explain the rise in power of information across the entire U.S. government, this lack of an overall theoretical construct ultimately endangers the overall stability of IO. This is because theory serves as a basis on which to build a model of a complex subject like IO so that it can be better understood. For example, an overarching academic theoretical construct on the order of realism or international liberalism, which can explain IO with sufficient rigor, does not presently exist. That is not to say that there have not been influential academics who have set forth theories for discussion and review. The books *Power in the Global Information Age: From Realism to Globalization* (Nye 2004) and *The Emergence of Noopolitik: Toward an American Information Strategy* (Arquilla and Ronfeldt 1999) are two examples; namely, they set forth the concepts of *Soft Power* and *Noopolitik*. However, there has not yet been an overwhelming acceptance of either of these constructs. For example, as part of the literature review in chapter 2, "A Theoretical Review of Information Operations in the United States," the arguments regarding soft power, as set forth in the seminal book, *Power and Interdependence*, are described in detail (Keohane and Nye 1989). These academics portray how the use of information is changing the idea of what is looked for in the power capabilities within the world political structure (Keohane and Nye 1989, 23). Robert Nye also captured the excitement and the power inherent in information in other books as well, such as *Bound to Lead*, and later amplified in other publications (Nye 1990; Nye and Owens 1996; Nye 2004; Nye 2006).

Yet, this research is really about a focus on power and its transformation as the world enters the Information Age. It is in this chaotic early stage of a new era when the disconnect between theory and reality is perhaps greatest; in the United States, the inability to match a strategic theory to the changes in the power structure of the federal government is most noticeable. So while *Soft Power* and *Noopolitik* may have struck a chord within the DOD and a number of federal agencies at some point, to date, none of these attempts to develop an overall encompassing IO academic theory for what is happening with regard to information has been formally adopted across the United States. Even Arquilla and Ronfeldt note as much in a recap to their book *The Promise of Noopolitik*, which was published eight years after the original publication of *The Emergence of Noopolitik: Toward an American Information Strategy* (Arquilla and Ronfeldt 2007). Their initial enthusiasm for this theoretical construct has been dampened considerably by the events of 9/11, Operation Enduring Freedom, and Operation Iraqi Freedom, as well as by the way the Internet and the intellectual community have evolved in the last decade. The hopeful optimism of the 1990s with regard to the World Wide Web and the Internet has dimmed, owing to the awful realization that given the power of information, many individuals and groups have instead used this new technology to their advantage, whether for their political, financial or social gain (Arquilla and Ronfeldt 2007). Likewise, Arquilla and Ronfeldt also admit in their postscript that the early promises of a global community have been overwhelmed by day-to-day events, which tend to mitigate the promise of revolutionary change. Although they still believe that *Noopolitik* is an idea for the future, and they remain optimistic, they are also dismayed by a number of trends that have effectively mitigated much of the promised potential of this theoretical construct:

- Notions like noopolitik are gaining credibility, but all too slowly.
- Soft power lies behind all such notions, but the concept needs further clarification.
- Activist NGOs representing global civil society are major practitioners of noopolitik, but the most effective may be the global network of jihadists.
- American public diplomacy would benefit from a course correction (Arquilla and Ronfeldt 2007).

So none of these concepts can be properly considered a rigorous academic theory on IO, but instead more of a series of ideas around similar topics that are attempting to define this radical change in power. All of these arguments are very interesting, because as represented in the interviewee data, changes are occurring slowly in the development of overall theoretical construct, definitions are not defined, and the federal government as a network is not as responsive as desired, specifically because the U.S. government public diplomacy efforts are considered insufficient. Perhaps an argument can be made that, in reality, a revolution in warfare is occurring with regard to IO, perhaps not at the rate initially desired—but at a more evolutionary pace.

In this vein, a thread has emerged from the participants' data that the reason that no overall IO theory has emerged is because IO is a concept that supports so many different and disparate academic areas, making it difficult to unify a community around a single concept. The sheer diverseness of this transforming idea is easily seen at IO conferences, where the hard and soft topics are instantly separated into separate streams and only rarely touch each other at the plenary sessions. Computer security, psychological operations, electronic warfare, public affairs, and the other portions of IO by themselves are all incredibly complex areas, and to find a single comprehensive academic theory that can encompass the use of these warfare areas and the others that comprise IO is incredibly difficult.

DOES MILITARY DOCTRINE EQUAL IO STRATEGIC THEORY?

While no overall academic theory has emerged to adequately explain the rising power of information, the same cannot be said for the avalanche of policy that has been promulgated by the DOD. Military doctrine is different than academic theory, but for the DOD, it serves much the same purpose—to ground the operational missions in a series of overlapping policy and strategy. IO doctrine is no different and was developed over a number of years as part of a maturation process of theory in the United States. The first of these policies, DODD TS3600.1, was published in 1992 and kept at the Top Secret level throughout its use, owing to the restrictive nature of its contents. So while this was an attempt to start a dialogue within the DOD on the new capability of information warfare, its security classification in general restrained a more rigorous doctrinal exchange. The need for a general theory or overall strategy to fit these revolutions in technology still existed, which prompted a new concept— command and control warfare. Officially released as a Chairman of the Joint Chiefs of Staff Memorandum of Policy 30, *Command and Control Warfare*, on March 8, 1993, this document laid out for the first time in an unclassified format the interaction of the previously mentioned disciplines, including electronic warfare, operations security, deception, and psychological operations, and was designed to give the U.S. war fighters the advantage in this new information environment. Interestingly enough, command and control warfare is a more restricted concept than information warfare, which means that the DOD backed down from its initial broader strategy that was published in 1992 with regard to information warfare; instead, it issued a more constrained policy in 1993. This change centered on those core disciplines that the U.S. military was most familiar with and had a greater history of use. This pattern was to be repeated again a decade later in 2003 with the publication of the *Information Operations Roadmap*.

During this period, IO doctrine also continued to be developed after the publication of the original command and control warfare doctrine in 1993. There was a concerted push for declassification and better understanding of these concepts within the DOD, which resulted in the publication of DODD S3600.1, *Information Operations*. By downgrading this document to the Secret level, the DOD opened IO to an even wider audience. In a related effort, the DSB also published its report "Information Warfare—Defense" in November 1996. Together these documents attempted to clarify the differences between the older doctrine, and introduced the concept of Computer Network Attack (CNA) as an IO capability. However, there were still questions regarding IO definitions and lexicon that would not be fully addressed until the release of the seminal publication, Joint Publication 3-13, *Joint Doctrine for Information Operations*, on October 9, 1998. It is in this document that the military released for the first time an unclassified document that widely disseminated the doctrinal principles involved in conducting IO. A key lesson learned from the release of this document was the realization that both the White House and DOD staff needed to understand that they needed better coordination. This is owing to the fact that IO efforts are often conducted long before the traditional beginning of active hostilities, so the Pentagon may not always have the lead in every operation. This early and sustained interaction between federal agencies within the executive branch has also brought about a renewed emphasis on the IO organizational structure. Chapter 2 of this book, "A Theoretical Review of Information Operations in the United States," is dedicated to the intricate and complicated relationships of the ever-evolving IO organizational structure.

In addition, following the release of Joint Publication 3-13 in 1998, new doctrine continued to be published. The *IO Roadmap* was released in a classified format in 2003. The Secretary of Defense's *IO Roadmap* was published five years after the release of Joint Publication 3-13 and was considered a major step forward in the development of this warfare area within the DOD. This is because of the cumulative efforts during this period of 1998 through 2003 to update and change the military's strategy on IO based upon real-world operations and missions conducted by the services around the world. In doing so, the *IO Roadmap* concentrated more on the traditional aspects of IO, including the old concept of command and control warfare. In many regards, it was seen as a revalidation of these concepts. Subjects such as perception management, strategic communications, public diplomacy, and influence campaigns were subsequently minimized in the *IO Roadmap*; instead, this document is a more tailored doctrine on IO. This latest policy in the form of the *IO*

Roadmap also chose to concentrate more on the traditional aspects of IO, including electronic warfare, psychological operations, and computer network operations, and does not try to control areas that the military did not control. This is because the *IO Roadmap* is an official DOD publication. In most aspects, it is probably the best official document that broadly defines the American military strategic policy, since it concentrates much more on the traditional aspects of IO. This document is also probably more representative of the manner in which the DOD operates, thus in effect, the *IO Roadmap* may have narrowed the gap between strategic theory and tactical IO operations by lowering the expectations of higher-level IO policy for the United States. This is a preliminary conclusion, but it will be interesting to see if the *IO Roadmap* leads to a greater understanding by the U.S. government of the overall power and capability of information as an element of power in this new era.

While the *IO Roadmap* could be considered a huge change because of its more narrow focus on the traditional areas of IO, it once again highlights the huge mismatch between the strategic transformational promise of IO doctrine and the operational reality of how the DOD tactically conducts information activities and campaigns. In reality, the *IO Roadmap* may be the best pragmatic solution for the conduct of IO by the military. The new Joint Doctrine for IO, Joint Publication 3-13, which was published in 2006, built on the changes inherent in the *IO Roadmap*; it is another major step forward because it marked the growing comfort level with the embedded role of IO within basic military strategy and operations. The year 2006 may also come to be seen as the period when every aspect of IO in the national power structure moved forward. The information assurance community also saw the publication of the National Infrastructure Protection Plan, while the strategic communications arena saw the development of a long-awaited draft strategy. These documents, along with the *IO Roadmap* and the new Joint Publication 3-13, provide the military-approved doctrine on which to base future IO plans and operations. The real question is whether this growing set of policy and guidance documents and proliferation of IO-related organizations indicates a greater understanding of the power and capability of IO specifically and information in general as an element of power in this new era by the U.S. government and its constituent elements.

The end of 2006 also saw the emergence of additional pieces of strategic guidance and policy, one from the DOD and one at the interagency level, which could show alignment with many of the major themes promulgated in this book.

Specifically, in September 2006, the DOD released the *Quadrennial Defense Review Execution Roadmap for Strategic Communication*, which briefly summarized the problem facing the DOD in this operational area and laid out fifty-five tasks intended to remedy those problems. Strategic communication was defined earlier in the *IO Roadmap* as "Focused US Government processes and efforts to understand and engage key audiences to create, strengthen, or preserve conditions favorable to advance national interests and objectives through the use of coordinated information, themes, plans, programs, and actions synchronized with other elements of national power" (U.S. Department of Defense 2003). This new approach and definition was significantly better than previous doctrine that emphasized the "transmission of themes and messages." The new view also recognized that if one hopes to have any likelihood of positively influencing an audience, the first step must be listening to and understanding that audience, and thus hopefully avoiding the widespread (and sometimes accurate) global perception that the United States is so busy talking that it can't afford the time and effort to listen. Likewise, the *IO Roadmap* also stated that the U.S. military is not "sufficiently organized, trained, or equipped" to engage in full-spectrum strategic communication, and "changes in the global information environment" require a more coordinated and integrated effort. It emphasized the importance of "credibility and trust," and noted that that all elements of the U.S. government share the responsibility for this (U.S. Department of Defense 2003). For not only is effective strategic communications a government-wide responsibility, the DOD is by no means the senior player in this effort, and it must support the efforts of the DOS to integrate these efforts. However, within the DOD, several key capabilities require improvement. Most of these capabilities fall within the umbrella of IO in some way, including public affairs, psychological operations, and defense support to public diplomacy. The DOD also defined three key objectives in this *IO Roadmap*. If met, these objectives would significantly improve its ability to conduct effective strategic communications. First, the DOD needs to institutionalize a process through which goals and objectives in this issue area can be embedded within the development and execution of plans across all operational levels. Next, the doctrine to clearly define the roles, responsibilities, and relationships for strategic communications and its constituent elements needs to be developed. Finally, the military departments, such as the Department of the Army, and combatant commands, like the Central Command, must be provided the means to organize, train, and equip capabilities for this (*Quadrennial Defense Review* 2006).

WHY IS THE STATE DEPARTMENT NOT ISSUING STRATEGIC GUIDANCE?

While the *Strategic Communications Roadmap* provided the DOD with authoritative guidance with which to shape capabilities and operations, the interagency organizations had no such guidance. However, there is hope that broader policy may be adopted eventually. In the second of the major IO federal policies that was released in 2006, Karen Hughes circulated for coordination a memo in October of that year titled *U.S. National Strategy for Public Diplomacy and Strategic Communication*, under her hand as chair of the Presidential Coordinating Committee for Public Diplomacy and Strategic Communications. This was a much longer and more strategic document that set forth three strategic imperatives to guide American public diplomacy and strategic communications programs. The first of these initiatives was stressing the importance of presenting a positive vision of hope and opportunity, which would be rooted in basic American values. Next was the need to isolate and undermine violent extremists. The final imperative was to nurture common interests and values while emphasizing those that cross cultures, borders, and creeds. The draft strategy then went on to identify critical influencers who are able to reach "strategic audiences" and "vulnerable populations." The plan also emphasized the need for interagency coordination, because every arm of the U.S. government has an urgent mission in this arena. Its "action plan" was based on these three strategic imperatives, and nearly 40 percent of the entire document was devoted to specific and detailed plans and proposals. Finally, the draft strategy also examined several critical elements of communication, such as broadcasting or public opinion analysis, that would be necessary supports for a successful strategy, and it emphasized the need to be accountable for operations and to gauge whether any specific plan or program was being successful (*U.S. National Strategy for Public Diplomacy and Strategic Communication* 2006).

This plan was broad and inclusive, a major step forward that went well beyond anything that had existed previously. One major improvement over earlier efforts was that the Presidential Coordinating Committee charged with developing this strategy was not co-chaired, as it had been in previous incarnations, and thus did not suffer from divided leadership. Instead, this interagency group was led only by Karen Hughes, one of the most influential members of the Bush administration. Her unique power stemmed from her key relationship with the president and her position as one of his key advisers, and her guidance always had an *ex cathedra* aspect to it. It was thought at the time that this initiative provided a unique window of opportunity in which real progress could be made before the pressures of the 2008 elections and the administration changeover in 2009. However, there were weaknesses in the plan, and the first of these was its insistent focus on the Muslim/Islamic world. While that is quite normal in one regard, especially in its connection to the GWOT, in other ways, that emphasis is unfortunate because there are other areas of the world, such as Latin America, Asia, and sub-Saharan Africa in which America needs to be fully engaged in support of vital national interests. Another area in which the plan was inadequate was the almost perfunctory section on resources. Instead of a powerful and compelling call for greatly increased resources with which to wage the war of ideas and a detailed explanation of how those resources would enable the United States to advance its interests, the strategy only provided a weak one-liner about the need for "increased support." This is a fatal flaw, especially in a fiscal environment in which every dollar has several worthwhile programs calling for it. Such a weak request has virtually no chance of actually gaining the needed resources, which, to date, has spelled a quick demise for this noble effort.

Is Information Operations the best term?

Information Operations may not be the best term; it is only the latest in a series of DOD names for a concept that has existed for over thirty years. The name is too limiting because it tends to be only associated with the military rather than the entire U.S. government. Variously called information operations, information warfare, command and control warfare, public diplomacy, international public information, psychological operations, perception management, net-centric warfare, netwar, soft power, noopolitik, and strategic communications, all of these terms are inadequate to explain the true breadth and depth of transformation of power across the international community. The capabilities of deception, psychological operations, and electronic warfare, which can shape and influence the information environment, have all existed as part of the military repertoire for a long time, but the umbrella term of IO is a relatively recent doctrinal definition, with much of the critical thinking beginning in the mid-1970s.

The demise of the Soviet threat to the United States in 1989 and the shift from bipolar to multipolar political scenarios seriously affected American force structure and military doctrine. This, combined with the huge technological changes that have evolved over the last twenty years in computers, software, telecommunications, and networks, has revolutionized the way the United States conducts military operations, and there has been a marked concentration on understanding the role of information in conflict. It was becoming increasingly clear during the late 1980s and early 1990s to the war fighters and policy makers in the Pentagon that the side that controlled and retained the ability to conduct information campaigns accurately, as well as the ability to manipulate, use, and disseminate information, was going to be victorious. Strategic planners at the Joint Chiefs of Staff began to think and write new strategy, most of which was highly classified, that would utilize information as a war-fighting tool. The evolution of these different IO terms is laid out in the next few sections.

PROBLEMS WITH THE USE OF IO AS A TERM

To begin with, the very term IO was a compromise from information warfare. The military understood information warfare to an extent, but just as quickly as that term started gaining acceptance over command and control warfare in the armed forces during the 1990s, the term "IO" was foisted on the DOD in 1998. The reason for this was to broaden acceptance of this new form of warfare across the federal government, where many agencies were anemic to the term "warfare" itself, and so new language was needed to soften the term and allow this warfare area to be utilized across the different federal interagency organizations. IO was adopted as a neutral label that could be used by all government agencies in the United States involved in these types of activities. The term IO ran into trouble right away because it included the older command and control warfare areas, such as operations security, psychological operations, and electronic warfare, with corollary functions like civil and public affairs. It is widely known that the psychological operations and public affairs communities are very separate and distinct areas that have disparate missions, which could make it unethical to work together. Huge discussions and debates were conducted on how to separate these two activities in an IO cell. No matter what was suggested, the idea that any public affairs official would ever be involved in any operations that conduct psychological operations, influence operations, or perception management type activities is counterintuitive to their whole mission, which in many cases spelled disaster from the beginning. A great example of this was mentioned earlier in this book with the demise of the OSI in February 2002, after the senior DOD Public Affairs officer, Tori Clarke, torpedoed the entire concept of this new organization. It is exactly this area of IO, namely, perception management or the newest term of strategic communications, which promises the most changes with regard to the power of information. The ability to use the latest technology to influence people around the world is the form and articulation of power and informational capabilities that grabs the attention of many proponents of IO. So the correct label is very important, and whatever is eventually settled upon, the taxonomy for this new set of tools is the crux of the potential power of IO. However, finding one term to cover all of these disparate activities is often very difficult and perhaps not always possible.

A Swedish information warfare academic once said, "While the activities gathered under the umbrella concept of IO are not new in themselves, they attempt to coordinate and integrate them into an overall strategy which utilizes the rapid advances in information and communications technology" (Riegert 2002, 79). For example, there are elements of destruction that are not part of an IO campaign; likewise, not every public affairs activity has to be tied to information operations. In reality, if done correctly, all elements and their components of national power can be integrated into a satisfactorily planned, designed, and executed strategy to allow the United States to attain its national security goals in the new millennium.

THE NEED FOR TAXONOMY

Labels are incredibly important. Portions of IO, such as psychological operations and electronic warfare, are distinctly military terms; yet they function very similarly to tasks like diplomatic information activities and worldwide communication efforts that are conducted routinely by other agencies in the United States. Thus we see the difficulties in determining what exactly IO means and why changing labels has occurred so much in these areas over the last two decades. For example, the term "command and control warfare" was routinely accepted by the DOD in the late 1980s and early 1990s. The focus was on nodes and connections, with an emphasis on physical items, such as network operations centers and transformers. This was a primary mission in the area of warfare that the DOD could and did excel in. Yet the evolution to a warfare area of title beyond the limitations of the command and control warfare label continues to vex the United States almost twenty years after the publication of the original DODD 3600 series in 1993. That is because the moment the military is moved beyond the traditional areas of operations security, electronic warfare, and psychological operations and begins to encompass mission areas that include components of influence operations and perception management, the services begin to have difficulties with the theoretical aspects of IO. The broadening of command and control warfare to information warfare was the next logical step in the mid-1990s as the revolution in military affairs was the rage. Policy was formulated that ultimately resulted in the seminal doctrinal statement of Joint Publication 3-13, *Information Operations*, in October 1998. This was supposed to be the preeminent manual on how to conduct missions in this new era where information reigns supreme. The problem was that this publication was not a how to manual; instead, it was an attempt to redefine how the military conducted operations—a reach for a "new" way of warfare. And with all things revolutionary, it was a bridge too far. The various military services had trouble trying to implement this new military strategy, as well as organizing, training, and equipping it. Funding was also crucial, because it was very hard to fund these nebulous concepts. All of these issues led to a realization that the original Joint Publication 3-13 reached too far in terms of military theory. Since then, there has been a concerted effort by the DOD to reign in IO policy and doctrine to mission areas that are more traditionally focused on the respective armed services. Combine these ideas with the lack of a proper definition and taxonomy for IO, and it is easy to see the problems in implementing IO across the U.S. government—the future of this transformational capability may never be fully realized.

Why is the top-down approach to IO not working in the U.S. government?

While these incredible changes in technology are drastically changing the role of information with respect to power, and many parts of the military and business communities have embraced these changes, it still appears that the executive branch and the DOS are still very slow to understand the power inherent in information. The lack of a set of coherent theories or overarching doctrine is creating a gap between the new changes that are occurring with the tactical agencies, while there is still a need for a basic understanding at a more strategic level. Yet the fundamental fact is that the growth of information technology has accelerated the process of transferring power down and away from a centralized authority and into the lower levels of an organization. This decentralization of power, command, control, and decision-making authority can be seen in many instances in new DOD weapon systems, such as the Future Combat System, where all army infantrymen will have more information at their disposal than could have been fathomed a mere decade or two ago. The same can be seen in the economic globalization efforts, where the market is truly worldwide and no longer is a business confined to a local geographic area. The Internet and World Wide Web have forever broken down these barriers to communication and information transfer, bringing the power to groups that formerly did not have access to these capabilities.

IS THE REVOLUTION IN MILITARY AFFAIRS / DIPLOMATIC AFFAIRS AN ANSWER?

Is the revolution in military affairs still a viable concept? How about the revolution in diplomatic affairs? Is the U.S. government really ready to radically change its organizational structure to conduct operations in the Information Age? Probably not, for while everyone understands that nation state to nation state communication will never again be limited to pin-striped diplomats, cables, message traffic, or official communiqués, it is not apparent from the data gained in this research that the radical leap needed to transform the DOD or the DOS is happening very quickly, especially in the areas of strategic communications and perception management. Unfortunately, it appears that the United States has been very slow to take advantage of this new technology and instead is relying on the tried and true communication apparatus that has been the backbone of public diplomacy for the last sixty years. The demise of the USIA and the incredible slowness of the DOS to properly absorb the public diplomacy community has also contributed immensely to this incredible gap in the strategic capability of the American government to adequately project its message to influence people around the world. The Clinton and second Bush administrations are to blame for this gap, because while they have repeatedly talked the talk about the need for a "beefed up" public diplomacy capability for the United States, their actions (or inactions in the case of lack of funding), have contributed significantly to the drastic decline in the ability of the DOS to project its message. In this gap, the NSC has tried to conduct public diplomacy or strategic communications campaigns, but finally gave up after the retirement of their main proponent in 2005. Indeed, the DOS is now quoted as stating that public diplomacy and strategic communications are no longer a capability of that office (Waller 2007a, 289).

Yet the diplomatic corps is also greatly at fault. It was a resentment by the traditional segment of this federal agency toward the independence of the USIA that allowed it to be absorbed in 1999. While there is "lip-service" paid to the development of a public diplomacy core in the "I" group, in discussions with a number of DOD professionals, it rapidly becomes apparent that public diplomacy is not considered a fast track to promotion. This apathetic attitude or indifference is telling in the staffing of public diplomacy positions, the funding of public diplomacy initiatives, and even in the leadership of public diplomacy within the DOS. The inability of the NSC and DOS to jointly lead a Presidential Coordinating Committee in this very area since 9/11 is also very telling of the importance that these organizations put into this capability. In addition, the protracted search for a leader of public diplomacy in the DOS is also problematic as well. First, it was Charlotte Beers, a Texas advertising executive who lasted less than eighteen months. She was followed by rumors of Margaret Tutweiller, a Bush administration speech writer. Then there was a gap of two years with an acting official until Karen Hughes took over the position in 2005 amid much fanfare and hype. However, she left Washington for good only two years later; some changes had been made, but there had been no continuity in the role. It is the belief of many of the interviewees for this book that the lack of a long-term, dedicated IO professional to coordinate this very important role has damaged the ability of the federal government immensely in this area.

WHY IS THE STATE DEPARTMENT FAILING IN ITS PUBLIC DIPLOMACY ROLE?
This is mainly because the public diplomacy community is still embracing antiquated tools to transmit its message. Comparatively, little effort is made to understand and use new avenues such as blogs, websites, Internet chat rooms, and instant messaging to pass information to all segments of society. There are a variety of reasons for this, but one could be the loss of control. The DOS has traditionally preferred to centralize the "message" that it promulgates to other nations, and thus the use of mediums that are under their centralized control, such as radio, television shows, and embassy visits. The problem is that these methods, while laudable, are not enough in today's technically savvy world. Adversaries and enemies of America are filling these other mediums with hatred, lies, and distortions of the truth that only serve to hurt the United States. A vacuum is abhorrent to nature; once it is discovered, it will be filled. That is exactly what is happening today with the effort by the American federal agencies to spread the word. Because the federal agencies are fighting with one hand tied behind their back owing to archaic laws and the unwillingness to use the latest technological assets, the United States is losing in this war of ideas. And it is a war, as stated by many IO experts (Waller 2007b). Our enemies understand that they cannot defeat the United States on a military or economic basis, but they can hobble this superpower with a well-run information campaign. It was done before in

Vietnam, and many of these same tactics are being utilized again, albeit with newer technology and communication paths.

So why is it that the DOD can trust the latest technology with its youngest recruits by the U.S. government, and yet the DOS is unwilling to trust the citizens of this great nation to spread the words and ideas of freedom and democracy. The awareness level of the intended recipients of these messages from the American sources are often extremely savvy and can understand nuances very well. That is the key to success in this war of words, and the United States must use all sources at its disposal to promulgate the message about freedom and democracy; only then will the tide will turn in this battle, which currently it appears that the coalition is losing. We talk about a revolution both in military and diplomatic affairs, which are really the use of information to transform these traditional forms of power; we talk about globalization, which is really the use of information to transform economic power; and yet we do not talk about a revolution in information affairs. Perhaps this is because information is still not viewed as a source of power but instead as an enabler—it is technology, specifically information technology, that is seen as driving the revolution in military affairs, revolution in diplomatic affairs, and globalization transformations. But is it not also information and the flow of information as well?

Why is there no rhyme or reason to the IO training and education curricula?

When evaluating the sheer multitude of IO courses run by the U.S. government, it is apparent that the major problem is that none of these courses have any standards or common learning objectives upon which to base their curriculum. These different classes are normally based on different theories (service and agency), different skill levels of users (beginner to advanced), different ranks or grades of the audience (enlisted to flag/general officer), as well as different foci (strategic, operational, and tactical). So it should not be surprising that there are over seventy IO courses being taught by a variety of U.S. government organizations and commands, all of which have little to no interaction or integration. For example, one cannot obtain IO training in one service and then serve in a joint organization without needing additional specialized training. Additionally, there are no common denominators or goals that translate well across the American armed forces with regard to IO training and educational requirements. These and other standardization issues have thwarted the U.S. government and academia in moving toward the development of curricula emphasizing the power of information in general and IO in particular.

CAN LESSONS BE LEARNED FROM THE INFORMATION ASSURANCE COMMUNITY?

The participants interviewed for this book have noted this dichotomy between the information assurance and IO communities, and they have presented this subject a number of times in conferences around the world as well as published their findings in a number of scholarly articles. One good example of this was a sponsored collaborative discussion session among British, American, and Australian academics and military officers at the 2nd Annual IO Conference hosted in London in July 2003. During this daylong session, a tremendous amount of energy and analysis was devoted toward finding a solution to help develop better access to IO training and education capabilities across the three nations. The figure below is a synopsis of those efforts, and it reiterates what the participants of this project have been advocating for a long time, namely, any curriculum developed must be based on open and accessible standards, and a Web- or Internet-based set of courseware is the best answer to deliver content globally.

Options for Improving IO Training and Education Goals

IO TRAINING AND EDUCATION GOALS	MEANS
Delivery of training must be cheap and fast	Internet
Access must be worldwide and standard	Portal
Clear, concise, authoritative and readable	Textbook
Information battlespace	COP
Planning tool/checklist	Excell App
Study real world operations	Case Studies
Common IO definition/language	Taxonomy
Change perceptions and generate interest	Exercises
Parallel play/multiple courses	Interfaces
Worldwide IO game	EverQuest
Standard IO training material	CD-ROMs
Training must be standardized	Qualifications
Red teaming must be incorporated	VA Teams

While this matrix is not the sole answer to the problem, the authors believe that it may help to act as a checklist or guide to focus attention on possible solutions to these IO education and training goals. However, there is still a gap between the large number of military-oriented IO courses and the study of this academic concentration by civilian universities.

ISSUES THAT STILL EXIST WITH DEVELOPING COMMONALITY WITH RESPECT TO THE IO TRAINING AND EDUCATION SITUATION

The dichotomy between increased emphasis by the U.S. military on the conduct of IO and the lack of corresponding academic programs within academia is not unprecedented. Early work at the National Information Assurance Training and Education Center to develop a set of standards led to several industry professional standards, the National Institute of Standards publication 800-16, and the Committee for National Security Systems' series of publications. These standards, developed by the NSA, are now widely recognized throughout the DOD and interagency organizations as the de facto baseline of tasks for information assurance across the federal bureaucracy. In addition, the Committee for National Security Systems series has become widely used in academia through NSA-sponsored information assurance programs and curriculum. Together these groups are a hub of information assurance activity in which a tremendous amount has occurred in the last two decades. An entire cadre of information assurance professionals has been trained and now occupy key, influential positions within the federal government as a result of the education that they received from these programs. The key component of this success has proven to be the development of the Committee for National Security Systems standards, which are grouped into six categories (4011 to 4016). Updated on a regular basis, these serve as a baseline for all the certifications and academic programs sponsored by the NSA and National Information Assurance Training and Education Program as well as by the information assurance academic community in general.

If the problem of developing academic interest in IO is to be solved, several steps are required. They can be modeled on the steps originally recommended for the information assurance discipline over a decade ago. The first is to build personnel capacity; if IO is to become a civilian academic field, one must have sufficient faculty. The main problem is that very few college professors are trained in IO in the United States. Currently, neither the computer science, information assurance, nor information systems programs in the United States would be able to adequately respond to the increased demand for IO courses. For the long term, it will be necessary to increase faculty in all areas of information technology, not just information assurance and IO. Current IO practitioners should be encouraged to enter the professoriate by creating academic positions for professionally qualified individuals. Currently in the United States, there are no comprehensive IO curricula or graduate programs in academia. Nascent Master's curricula are underway at NDU and Norwich University, but more institutes and programs are needed to help close this gap if significant progress is to be made. Likewise, the role of industry also cannot be overlooked in making faculty retention and development easier for the IO initiative. It is also imperative to attract quality students to programs producing IO specialists. As demonstrated in various information assurance initiatives, an undergraduate scholarship program has the largest potential influence to solve the short-term problem. In the absence of some form of graduate stipend program, there will probably still continue to be a dearth of individuals who will become the next generation professoriate and fill governmental and industrial needs. Production of Master's and Doctoral students is also essential. Finally, traditional undergraduate and graduate programs alone cannot meet the need for information operations professionals, and any comprehensive solution must include ongoing professional education for the existing workforce.

CHAPTER 6

KEY FINDINGS IN THIS BOOK

Information has always been an element of power, but it is often seen as an enabler or supporting component and not as the decisive factor in conducting operations. The very nature of modern day operations, with its persuasive 24/7 global media coverage, has proven the need to utilize all the tools or elements of power. Information is a key component of any sort of influence type of operations, and its effectiveness has been demonstrated repeatedly, especially over the last two decades with the rise in technological capabilities. However, the very factors that make information so useful as an element of power are also adding to the difficulties for nation-states, and in this case, America, to conduct information campaigns, or IO, on a successful basis. The shifting of power away from a centralized authority, the loss of control from the federal bureaucracy, and the low cost as well as ease of entry into this domain all have combined to signal a revolutionary and radical shift in the manner that information is utilized around the world. Therefore, it is not surprising that NGOs, non-state actors, corporations, terrorists, and individuals have all benefited from this shift in power, owing to the advent of new information technology capabilities.

It is also not surprising that the federal bureaucracy of the United States is struggling to come to grips with the ramification of these changes. Specifically, the loss of control, flow, content, and communication paths of information have all radically altered the method in which the administration and other branches of the federal government interact with their counterparts around the world. Combined with the heightened expectations of the increased capabilities inherent in IO; the lack of a coherent theoretical construct, definition, or taxonomy; and a virtual smorgasbord of training classes that have varying curriculum and content, none of which are integrated or coordinated, spell disaster for the success of IO in the United States. Too much is expected, and too much has been promised, and with no radical changes in funding across the federal agencies, progress has been disappointing. Many of the same organizations that were doing command and control warfare over fifteen years ago are still the key agencies conducting IO. They have been renamed and slightly expanded, but they have no true increase in scope and capability. Therefore, it is not surprising that in many aspects, some consider the DOD to be moving backward with regard to strategy, capabilities, and scope. The inability of the military forces to organize, train, and equip the nebulous original Joint Publication 3-13 directive of 1998 pushed the DOD to revert to the instruction of the *IO Roadmap* in 2003, which more closely resembles the original command and control warfare doctrine of ten years earlier. This was because it is precisely these latter capabilities, incorporated in electronic warfare, deception, operations security, psychological operations, and physical destruction, that the military has total control over, as opposed to more nebulous IO warfare areas, such as perception management and strategic communications. This "boxing in" of the DOD is actually a sound strategy, because it concentrates on what units and personnel are under its control and success in these areas. Taken together the specific key findings that align with this assessment include the following:

- IO needs to be limited in its scope to be effective—a lessening of expectations.
- Only the DOD will continue to have IO policy and doctrine.
- IO training and education are useless unless tied to taxonomy and standards.
- IO metrics are key to future success and acceptance.

All of these issues will be addressed below as part of an overall plan to provide a way forward with regard to the more efficient conduct of IO by the U.S. government.

Suggestions for improvement based on the soft system methodology and literature review

From this analysis, a number of specific recommendations were made that were both feasible and desirable. These suggestions are listed over the next few pages and represent the collection of several years of interviews, conferences, and workshops in an attempt to ensure that the specific recommendations of this book met all of the criteria of the participants. For as many academics have tried to articulate, this new emphasis on the use of information is an attempt by the United States to develop a strategy to better control all of its power capabilities in order to affect the many issues that it must deal with in the post–Cold War era. Federal officials in the United States have come to the realization that militarily, the government could not solve all of its problems through kinetic means. Therefore, IO is an attempt to bring these different facets of power to bear on an adversary, whether it is a nation-state, terrorist group, or individual. Thus the real key to making the management of information effective is to ensure that the horizontal integration and coordination of the interagency organizations are conducted in the peacetime environment versus waiting until hostilities start. As mentioned earlier, IO can be a very effective tool for shaping the environment in the pre–hostilities phase; sometimes, the need for hostilities may actually be avoided or minimized. So while the publication of Joint Publication 3-13 was lauded in the late 1990s with its attempt to define everything as IO, its very overstretch could actually be responsible for the lack of understanding and the eventual withdrawal of this strategy into a more manageable set of IO doctrine five years later with the promulgation of the *IO Roadmap*.

This latter argument is a key point of this book, namely, in trying to be everything for everybody, IO has, in effect, became nothing. It was emphasized repeatedly from the book participants that in order to be successful, IO needs to be more strictly defined and standardized with a combination of overarching policy, taxonomy, certification, and methodologies that are recognized and understood by all practitioners. In order to do this, many interviewees recommended that the IO definition must be limited to a more realistic view, in which goals and capabilities are attainable. It was stated repeatedly by the participants in this research project that nebulous policy and the desire to include all warfare areas into IO have actually hobbled the ability of the U.S. government to organize, train, and equip its forces in a practical manner to conduct these operations.

A PLAN TO DEVELOP AN OVERARCHING IO THEORY

IO is not a part of the liberalism or realism theoretical academic theories. It is something that is in between because it is much more oriented around power. It has its own language, such as viruses and worms, that is somewhat medical in nature. It also can be very technical, especially when concerned with information assurance or cyber security issues. This dichotomy of needs and requirements has hampered the ability to develop an overarching IO theoretical construct, and yet many comments from the participants interviewed for this book centered around the desire for more progress to occur, especially in the areas of IO standards, training, and integration. The use of IO policy and themes are very different across the U.S. government, particularly in the perception management arena, while computer network defense and critical infrastructure protection are considered more uniform in nature. Concerns were raised about why IO is so easy to visualize and so hard to accomplish. It is the softer areas of IO—the concepts that involved efforts to affect the mind in the form of perception management and strategic communications—that the United States was having the most difficulty conducting operations in. These skill sets are considered an art; many of the interviewees believe that one needs to take the long view for success in this area. Yet these same participants also noted that in the United States, federal organizations often wanted to rely on technology to answer the questions, and to overwhelm the adversary quickly. These interviewees commented that in reality, the fast results are not successful; instead, one should look at history to understand that quick answers are not the norm and how the military actors in the past have really succeeded. For example, was the bombing really effective in Kosovo in 1999, or was the NATO coalition just making rubble bounce without truly understanding how to affect the hearts and minds of a populace? For many of the research interviewees, IO is not that radical. Instead, it should just be labeled as "Operations in the Information Age." But that idea doesn't solve the need for an overarching construct, and developing new academic theory often hinges on radical concepts, such as those espoused in the *The Third Wave* or *The Emergence of Noopolitik: Toward an American Information Strategy* (Toffler 1984; Arquilla and Ronfeldt 1999). These concepts, along with soft power, are perhaps the best examples of academics that have successfully crossed the theoretical construct boundary into DOD policy (Nye 1990).

In this aspect, there is a huge dichotomy in the goals of these two policy attempts at developing strategic IO academic theory, with the more pragmatic DOD (*The IO Roadmap)* and the DOS *(Defense Science Board Report on Strategic Communications)*, documents. But in another view, these mandates are also entirely representative of the way in which IO is conducted today throughout the federal bureaucracy. Because the *IO Roadmap* has a much narrower focus than the mandate from the Defense Science

Board, it tends to highlight the huge mismatch between the strategic transformational promise of IO doctrine with the operational reality of how the DOD tactically conducts information activities and campaigns. So in reality, the *IO Roadmap* may very well be a pragmatic solution to the difficulties in trying to conduct these types of information campaigns on a day-to-day basis. The lofty and somewhat more ambitious goals of the DSB report, which while utterly correct from a perception management perspective, may never occur owing to political and fiscal reality.

The *IO Roadmap*, and to a lesser extent, the new, 2006 Joint Publication 3-13, are not the only way ahead for the federal bureaucracy with respect to the future of IO within the United States. In September 2004, a new DSB Task Force released the *Report on Strategic Communications* as a follow-on to an earlier study by the DSB in October 2001. Many critics felt the first study was overshadowed by the tragic events of 9/11 and the opening campaign of Operation Enduring Freedom in Afghanistan. So a primary duty of this new *Report on Strategic Communications* was to look at the changes that had occurred since the original report and to reflect on the prior publication to see if its recommendations were still valid. According to the report's opening statement,

> This Task Force concludes that U.S. strategic communication must be transformed. America's negative image in world opinion and diminished ability to persuade are consequences of factors other than failure to implement communications strategies. Interests collide. Leadership counts. Policies matter. Mistakes dismay our friends and provide enemies with unintentional assistance. Strategic communication is not the problem, but it is a problem (Defense Science Board Task Force of the Report on Strategic Communications 2004, 1).

The report went on to cite seven key factors for success with regard to strategic communications by the United States. All of these areas were important, but without an administration and federal bureaucracy that understand the problem, give strong leadership, and encourage a strong government-private sector partnership, this DSB report saw little chance of success for strategic communications, notwithstanding its recommendations, which are laid out below:

■   Issue a National Policy Security Directive on strategic communications from the NSC.

- Establish a permanent strategic communication structure within the NSC to include a Deputy National Security Adviser and a Strategic Communication Committee.
- Create an independent, nonprofit, and nonpartisan Center for Strategic Communication to support the NSC.
- Redefine the role and responsibility of the Under Secretary of State for Public Diplomacy and Public Affairs to be both policy adviser and manager for public diplomacy.
- The public diplomacy office directors in the DOS should be at the level of deputy assistant secretary or senior adviser to the Assistant Secretary.
- The Under Secretary of Defense for Policy should act as the DOD focal point for strategic communication.
- The Under Secretary of Defense for Policy and the Joint Command Staff should ensure that all military plans and operations have appropriate strategic communication components. (Defense Science Board Task Force of the Report on Strategic Communications 2004, 10).

What is very interesting from an academic standpoint is that many of the personnel interviewed for this DSB project also participated in this book, and many of the recommendations of this report mirror the overall tone of this book. In addition, all the key interviewees of the DSB previously worked or currently work with the public diplomacy, strategic communications, or international public information community, which validated their findings. Therefore, in a manner, this DSB report also serves as a verification of sorts with respect to the research conducted as part of this book to confirm that the assumptions are on track with regard to the needs and deficiencies of IO within the U.S. government.

Thus the way ahead for developing a strategic IO theory will have to involve the academic community. Unfortunately, as mentioned earlier, there are very few American university professors who have expressed interest or expertise in IO, so the ability to house this effort in a U.S.-based academic venue is unlikely. Fortunately, an academic IO theory does not have to be developed by an American to be useful. A tremendous amount of talented and innovative research on IO is being conducted outside of the United States, and so a collaborative approach is suggested with the three main IO and information warfare academic conferences are utilized as the backbone for this effort. Titled the European Conference on IW, the Australian

IW Conference, and the International Conference on IW, these three gatherings are held yearly and typically have many of the same participants attend from around the world. This makes for a nice setting for a vigorous debate, one in which a number of aspects and options to developing a strategic IO theoretical construct are analyzed with sufficient academic rigor.

A MODEL TO ESTABLISH A TAXONOMY AND SET OF DEFINITIONS FOR IO

Ultimately, the lack of a standardized nomenclature or taxonomy hurts the ability of the U.S. government to conduct IO. Basic questions are raised, including those of a semantic nature, such as why couldn't other U.S. government agencies agree on a common taxonomy, or a set of terms, such as information warfare? Was it too war-like, hence the switch to IO? Maybe so, but even the latter term is still not routinely adopted across the federal bureaucracy, and there are neither common terms in other organizations for IO nor terms for its different sub-themes such as perception management, international public information, public affairs, and strategic communications.

So it is suggested as part of this book that a set of definitive definitions and taxonomy needs to be developed to support the entire federal bureaucracy with regard to IO. A top-down approach has been suggested as part of the interviewee process, and the use of the three main IO and information warfare academic conferences could offer a way forward in solving this issue. Specifically, at these latter venues, streams should be set up to develop an ontology and knowledge base of IO, based on the author's role on the editorial staff of these conference committees. Ontology is a hierarchy of what is known and understand about a subject. A knowledge base is a web of relationships among the items in the ontology. This web of items and how they are related defines this knowledge base. As part of this effort, it is suggested from the interviewees that a portal should be developed—or at the least, a web-based service—that academics can use to access this knowledge base. It is proposed that the following items will also need to be addressed at these academic conferences, and then be included in this web application as it is developed:

- A clear definition of what IO is and how it works
- A glossary of IO, information assurance, information warfare, and other terms
- A mind map of important things related to IO and how they are related in multiple

ways (A discussion on these relationships' connections in the mind map indicates a variety of relationships among the items on the mind map.)

- A list that is part of and is contained by a breakdown of IO components—that is, methods, processes, who uses them, what they are, how they relate to action states—offense, defense, and collaboration
- A mental model of how IO is used, by whom, where appropriate, and all players, info, data, and knowledge common among them

AN ANALYSIS OF WHICH APPROACHES AND PROCESSES
WORK BEST TO SUPPORT IO

What all these policy developments and organizational changes have recommended and attempted to explain is a much greater emphasis on the use of the information environment across the spectrum of national security activities, from perception management capabilities by the federal bureaucracy to engage in strategic operations in the GWOT to securing critical information infrastructures against terrorist attack to military employment of the full range of IO's core competencies. The participants in this book were also very vocal and adamant in their desire for changes to be made in the conduct of IO by the U.S. government. Most of these can be grouped into the offensive IO category. Questions were asked repeatedly, such as, "Can offensive IO succeed?" "Should we try to do offensive IO?" "Does offensive perception management work better when done naturally?" In addition, some participants noted that more emphasis should be placed on the publicizing of key American documents in IO missions, such as the Declaration of Independence, the Constitution, and the Bill of Rights. The key to success in offensive perception management, as opined by a number of participants, is to keep it simple, to use a small number of common themes and goals that recap the lessons learned over and over, and to do it across the entire federal government consistently. In order to succeed, these same participants also noted, one needed to get top-level government officials to buy in and then go out and preach as well teach at all levels, with freedom and democracy as constant themes. Success in this kind of approach was considered more of a long-term goal, not something that can be considered an overnight success. A good example of this kind of methodology was the USIA, which at its peak concentrated on the economic, social, and diplomatic areas in its effort to combat communism rather than the military missions. These efforts were considered to be huge successes with regard to perception management, where the federal government let organizations other than the DOD lead the effort.

The research for this book was conducted over a long time period. Preliminary research began before the horrific events of 9/11 and continued throughout the Operation Enduring Freedom and Operation Iraqi Freedom campaigns. Early on, the research focused more on computer network attack, computer network defense, and critical infrastructure protection, all of which are more computer-centric issues that were considered key because of the enormous changes that were foreseen with the rise of the Internet at that time. As the research continued, it became clear that while the information assurance, computer network defense, and critical infrastructure protection issues are still vital areas in which to conduct research, they are all under some sort of control. There are organizations in the U.S. government around which IO policy and personnel are in place to handle or coordinate many of these defensive issues, and while these areas may not be totally solved, at least to some extent there are a series of standard operating procedures, methodologies, and processes at work. The same cannot be said for IO issue areas, such as perception management and strategic communications. Therefore, the thrust of this research is also that different methods work well for different parts of IO—namely, that a top-down approach on defensive options in the computer network defense and critical infrastructure protection areas may work better than a bottom-up approach that centers around perception management and strategic communications. Overall, the book participants agreed that one methodology is not the best for all areas of IO. There were many reasons for this, but perhaps the easiest to explain is that because IO is such a complex operational area, combining multiple diverse and time-honored warfare areas with new and complex capabilities is difficult because the former all have their own traditions and histories. In order to lay IO into this mix as an umbrella type concept, a varied methodology approach is more likely to succeed over one single approach.. Therefore, in order for the U.S. government to continue moving ahead with respect to IO, it is suggested per the interviewee data that a combination of techniques, methodologies, and processes must be utilized.

If the new Joint Publication 3-13 and the *IO Roadmap* are now considered the preeminent DOD policies on the power of information, it has to be questioned whether they really are the ultimate solution to the problems affecting the federal government with regard to the operational capabilities of IO, or are they a series of compromises by the military services and an attempt to publish a more "realistic" answer to "operationalizing" IO across the DOD? This narrowing of the IO policy is in opposition to what many of the interviewees recommended, for as noted throughout this section as well as in a large number of documents in the literature review, a much greater emphasis on the use of perception management capabilities by the federal bureaucracy was suggested to engage in strategic operations in the GWOT. The *IO Roadmap*, which was promulgated by the DOD in 2003, does not appear to follow these recommendations as suggested by the book participants; instead, it appears to consolidate IO into more discrete military warfare areas and is more aligned to the older command and control warfare policy. Thus the recommendation for this key theme of this research is to fund and promote understanding of where the true changes in IO will probably come in a decentralized manner, or as one interviewee stated, "change occurs best at the edges" (Rendon 2003). Opportunities to evolve policy, organizations, training, and tools in small but significant areas should be viewed as a good approach to follow for the conduct of IO across the federal bureaucracy, with the understanding that they offer the most hope to eventually produce the revolutionary effects that were envisioned from IO nearly fifteen years ago.

DEVELOP AN INTERNATIONAL STANDARDS EFFORT WITH RESPECT TO IO TRAINING AND EDUCATION

Based on the interviewee data, a suggestion has arisen that involves the establishment of an internationally based IO Working Group (IOWG) to conduct the following activities:

- Create the IO Working Group manifesto
- Create relationships with the police, the military, professional bodies, other defense agencies, and the corporate world in the participating countries
- Coordinate a series of International Information Operations Standards for IO workshops
- Develop and publish IO standards

Specifically, after a recent International Conference on Information Warfare that was held at the Naval Postgraduate School in Monterey, California, in March 2007, the following deficiencies in IO were identified:

- IO is a field that has no current standards.

- After the recent technological developments, the stakeholders of IO are not just nation-states and military groups any more; they are commercial and governmental organizations that are members of the critical national infrastructure of a nation.

- IO is a multifaceted discipline that brings together specialists in computer science, sociology, psychology, communications, international relations, and military science.

- There is a need for the aforementioned parties to be able to cooperate and collaborate for producing standards and defining the science of IO.

The first step to mitigate these issues is the proposed creation of a virtual community that would bring together the members of the working group for identifying and producing a course of action. It is suggested that this steering group will utilize a website, create a series of mailing lists, and use existing scientific conferences to disseminate results. The steering committee of the IOWG will be expected to promote the principles of IO in their respective countries and identify and establish relationships with stakeholders: academia, professional bodies, the corporate world, the military forces, other defense agencies, and law enforcement. This involves organizing a series of meetings, organizing workshops, and disseminating results following traditional publication approaches. At this stage, it is considered that one annual workshop will be adequate.

The second milestone is the development of the group's manifesto. Once the steering committee of the IOWG is established, it will produce a manifesto, and the future actions of the IOWG will be dictated by it. The group will develop a collaborative set of IO standards that will be disseminated via journal papers, conferences, workshops, and press releases. The third milestone is the creation of relationships with the European Network and Information Security Agency, the U.S. DOD, the United Kingdom Ministry of Defense, the Finnish Ministry of Defense, and the Research Network for a Secure Australia. Ultimately, the main outcome will be the creation of international IO standards that will be released. It is possible there will be two sets of standards, one for military operations and one for the public. It is hoped that the establishment of this IOWG will greatly improve the capabilities of a set of IO standards, especially across the United States and its federal bureaucracy.

Developing standards alone will not meet all of the needs for IO training, and there is no fast and simple solution. By encouraging and increasing the capacity of current programs, there will be an immediate, small increase in flow created by accelerating the progress of students currently in the programs. Currently, the production has been increased to a few hundred a year. Experience with the Information Assurance Scholarship Program indicates that de novo programs take as long as four to five years to produce the first individuals with baccalaureate degrees focusing on IO. To produce individuals at the Master's level takes an additional year and a half, and then an additional two to three years to produce a Ph.D. The foregoing discussion provides investment solutions that initiate and rapidly build an IO educational infrastructure for the long-term national interest. It involves:

- Investing in undergraduate and graduate students to encourage them to enter the profession
- Investing in current faculty to keep them in academia
- Investing in converting faculty to support IO initiatives
- Investing in research to maintain the state of the art and advance the profession
- Aiding in the development of information operations as a recognized discipline in conjunction with information assurance
- Aiding faculty in professional development and publication of research results

The following nine-point program would establish an integrated academic infrastructure dedicated to providing the education and training required to support using IO to protect elements of the critical national information infrastructure. Specific actions proposed include:

- Creation of a scholarship program to encourage both undergraduate and graduate students to enter the profession
- Creation of distinguished professorships and associated stipends to encourage faculty both to join and to remain in the academic ranks
- Creation of joint research opportunities with the government
- Creation of mechanisms to maintain currency of teaching and research facilities
- Encouragement of government, industry, and academic personnel interchanges
- Encouragement of joint academic–industry research consortia to address current needs

- Creation of an information operations training program to increase the number of faculty teaching and researching in the area
- Creation of joint education and training programs to keep current practitioners current
- Encouragement of the creation of innovative research outlets for faculty

The emphasis of this push to upgrade the IO training and education curricula is to help support the attraction of qualified personnel and students to the profession, with the development of a sufficiently large and well-informed faculty to guide education, training, and research programs for these personnel and students. In addition, improved infrastructure is needed to support such programs, as well as strengthening ties between industry, government, and academia through joint education, training, and research initiatives and opportunities. Finally, as has been emphasized in Estonia and Georgia recently, the use of cyber warfare tactics is becoming more prevalent. Training and education in information assurance and IO capabilities, with the development of appropriate standards, could also help to alleviate some of these risks and vulnerabilities.

Areas for Future Research

All of the areas addressed above are considered to be key findings, and if extended, they could also be logical areas to conduct additional research in the future, with specific focus areas to include the following suggested topics:

- The reasons why the DOD limited its IO policy
- The reasons for the lack of a strategic academic IO theory
- Research attempts to link IO training and education to taxonomy and standards
- The use and success of metrics in IO missions

In addition, since this research was primarily conducted in the United States, opportunities exist to research similar test cases outside of America. Likewise, this book also emphasized the fact that no longer will large organizations be the appropriate entities to execute this element of power; instead, it will be the nimble and smaller activities and agencies that succeed in this new era. Future research could be conducted on the optimal size of an agency or group that is best suited to this new informational environment. Also, other academic issues that are available for research could revolve around which organizational structure can be used to best maximize its capabilities in the Information Age, whether it is at the strategic, operational, or tactical level. In addition, more research could be conducted in the key features that were mentioned in the first chapter section, namely, open communication links, little censorship, truthfulness of information, and whether strengthening networks will decide the future of the world's political structure. Finally, in this book, definitions and models were developed that articulate not only why this divide between strategic theory and tactical operational missions exists, but also specific strategies for utilizing IO in a manner that best optimizes the inherent capabilities of this element of power. Taken together, all of these areas mentioned above could be lucrative for research by academics in the future as a result of the incredible change that is occurring within this issue area.

Conclusion

In summary, what all of the interviewees emphasize and acknowledge, which is also alluded to in the books, articles, conferences, and reports that make up this book, is that a drastic change in the conduct and use of power has occurred during the Information Age. In this book, these changes were discussed with a large number of personnel as part of this research and focused not only on the changing nature of power with respect to information, but also on the growing power of information itself. In addition, the author has attempted to analyze how these recommendations from the data gathered in this project compare to the actual development of IO by the U.S. government. Likewise, this research also compared the changes recommended in the conceptual models to other literature on this subject in order to analyze whether other authors agreed with the research participants as the way ahead to further the progress of IO. With regard to the literature itself, some of these books and articles were prescient and seminal while others were less useful and have quickly faded into obscurity. There are many reasons for this, but as the author's hypothesis suggests, IO policy often does not readily translate into the tactical operations. Therefore, what the literature review noted earlier, "A Theoretical Review of Information Operations in the United States," and its analysis have attempted to do is reiterate and show the gaps in literature between strategic doctrine and the day-to-day reality of this new warfare area, as well as how this research intends to fill that void. Finally, the author also attempted from the aspect of this research to show the gap in knowledge that exists today, not only from a literature analysis perspective, but also by comparing it to the requirements for the continuing development of IO, with an extensive series of interviews over a ten-year period.

In conclusion, what all of these texts and the book interviewees recognize is that there is a new role for information as an element of power. This author understands that it is the fungibility of information that makes it so truly useful and has attempted to emphasize that the ability to transform information, and to move it or display its capability, all relates directly to its power. This is the concept of strategic IO that quickly captures the minds of so many because of its great potential. Many of these texts also point to a more realistic appraisal of the current IO capabilities of the U.S. government. They often suggest that the best way forward is a more pragmatic approach of continuation and maturation of the IO process within the federal bureaucracy. Therefore, the challenge of this research has been an attempt to analyze the gap between the strategic concepts of the power of information envisioned by the United States and the way IO is actually conducted by the government in order to help formulate a plan to lessen this gap, based on the suggestions of the interviewees.

Bibliography

Ackoff, R. L. *Creating the Corporate Future*. Chichester, UK: Wiley, 1981.

Adams, J. *The Next World War: Computers are the Weapons and the Front Line is Everywhere*. New York: Simon & Schuster, 1998.

———. Interview with author. June 5, 2000.

Advisory Group on Public Diplomacy for the Arab and Muslim World. *Changing Minds, Winning Peace: A New Strategic Direction for U.S. Public Diplomacy in the Arab and Muslim World*. October 1, 2003.

Alberts, D. S. *Defensive Information Warfare*. Washington, DC: Institute for National Strategic Studies, National Defense University, 1996.

———. *The Unintended Consequences of Information Age Technologies*. Washington, DC: Institute for National Strategic Studies, National Defense University, 1996 and 2002.

Alberts, D. S., and Daniel S. Popp, eds. *The Information Age: An Anthology on Its Impacts and Consequences*. Washington: Institute for National Strategic Studies, National Defense University, 1997.

Albright, M. *The Importance of Public Diplomacy to American Foreign Policy*. Washington, DC: U.S. Department of State Dispatch, October 1999.

Allison, G. T. *Essence of Decision*. New York: HarperCollins Publishers, 1971.

Alter, J. "Why Hughes Shouldn't Go." *Newsweek*. May 6, 2002.

Anderson, R. H., P. S. Antón, S. C. Bankes, T. K. Bikson, J. Caulkins, P. J. Denning, J. A. Dewar, R. O. Hundley, and C. R. Neu. *The Global Course of the Information Revolution*. Santa Monica, CA: RAND, 2000.

Arkin, W. "The Other Kosovo War: Baby Steps—and missteps—for information warfare," August 29, 2001, http://www.msnbc.com/news/607032.asp?cp1=1.

Armistead, E. L. "Fall from Glory: The Demise of the USIA during the Clinton Administration." *Journal of Information Warfare* 3, Vol. 1 (May 2002).

———. "Back to the Future: Strategic Communication Efforts in the Bush Administration." *Journal of Information Warfare* 3, Vol. 2 (Fall 2003).

———. ed. *Information Operations: Warfare and the Hard Reality of Soft Power.* Washington, DC: Potomac Books, Inc., 2004.

———. ed. *Information Warfare: Separating Hype from Reality.* Washington, DC: Potomac Books, Inc., 2007.

———. "A Tale of Two Cities: Approaches to Counter-Terrorism and Critical Infrastructure Protection in Washington, DC and Canberra." *Journal of Information Warfare* 1, Vol. 3 (Spring 2004).

Arquilla, J. "The Velvet Revolution in Military Affairs." *World Policy Journal* (Winter 1997–1998).

———. Interview with author. February 19, 2003.

Arquilla, J., and D. Ronfeldt. *CyberWar is Coming!* Santa Monica, CA: RAND, 1992.

———. "CyberWar is Coming!" *Comparative Strategy* 2, Vol. 12 (Summer 1993): 141–165.

———. *The Advent of Netwar.* Santa Monica, CA: RAND, 1996.

———. eds. "A New Epoch—and Spectrum—of Conflict." In *Athena's Camp: Preparing for Conflict in the Information Age*, 1–22. Santa Monica, CA: RAND, 1997.

———. eds. "Looking Ahead: Preparing for Information-Age Conflict." In *Athena's Camp: Preparing for Conflict in the Information Age*, 439–501. Santa Monica, CA: RAND, 1997.

———. *The Emergence of Noopolitik: Toward an American Information Strategy.* Santa Monica, CA: RAND, 1999.

———. *Swarming and the Future of Conflict.* Santa Monica, CA: RAND, 2000.

———. *Networks and Netwars.* Santa Monica, CA: RAND, 2001.

———. "The Promise of Noopolitik." *First Monday* (July 20, 2007).

Ashby, William Ross. *An Introduction to Cybernetics*, 202–218. London: Chapman and Hall, Ltd., 1956.

Avison, D., C. Cuthbertson, and P. Powell. "The paradox of information systems: strategic value and low status." *Journal of Strategic Information Systems* (1999): 419–445.

Avruch, K., J. L. Narel, and P. C. Siegel. *Information Campaign for Peace Operations.* Washington, DC: C4ISR CRP, 1999.

Babbie, E. *The Practice of Social Research.* Belmont, CA: Wadsworth/Thomson Learning, 2000.

Baldwin, D. "Interdependence and Power: A Conceptual Analysis." *International Organization* 4, Vol. 34 (Autumn 1980).

Barber, B. "Group Will Battle Propaganda Abroad; Intends to Gain Foreign Support for U.S." *Washington Times.* July 29, 1999.

Barnett, F. R., and Lord Carnes, eds. *Political Warfare and Psychological Operations: Rethinking the U.S. Approach.* Washington, DC: National Defense University Press in cooperation with National Strategy Information Center, 1989.

Barnett, R. W. "Information Operations, Deterrence and the Use of Force." Newport, RI: Naval War College Review, 1998.

Bass, C. D. *Building Castles on Sand? Ignoring the Riptide of Information Operations.* Montgomery, AL: Air War College, 1998.

Beetham, D. "The Legitimation of Power." *International Relations* 56, Vol. 3 (1991).

Bennett, J., and C. Munoz, "USAF Sets Up First Cyberspace Command." *Inside Defense.Com* (November 4, 2006).

Bennis, P., and M. Moushabeck, eds. *Beyond the Storm: A Gulf Crisis Reader.* New York: Olive Branch Press, 1991.

Bering, H. "Professor Albright Goes Live." *Washington Times.* August 4, 1999.

Bernhard, N. "Clearer than Truth: Public Affairs Television and the State Department's Domestic Information Campaigns, 1947–1952." *Diplomatic History* (Fall 1997).

Blackburn, P. P. "The Post Cold War Public Diplomacy of the United States." *Washington Quarterly* (Winter 1992).

Blackington, R. "Air Force Information Operations (IO) Doctrine: Consistent with Joint Doctrine?" Research report submitted to the faculty of the Air Command and Staff College, Air University, April 2001.

Biocca, F., and M. R. Levy, eds. *Communication in the Age of Virtual Reality.* Hillsdale, NJ: Erlbaum, 1995.

Bresheeth, H., and N. Yuval-Davis, eds. *The Gulf War and the New World Order.* London: Zed Books Ltd., 1991.

Brocklesby, J., and S. Cummings. "Foucault Plays Habermas: An Alternative Philosophical Underpinning for Critical Systems Thinking." *Journal of the Operational Research Society* 47 (1996): 741.

Brookings Institution. *Financing America's Leadership*. Report of an Independent Task Force sponsored by the Brookings Institution and the Council on Foreign Relations. New York: Council on Foreign Relations, 1997.

Brown, R. "Information Operations, Public Diplomacy and Spin: The United States and the Politics of Perception Management." *Journal of Information Warfare* (May 2002).

Bruner, J. *Acts of Meaning*. Boston: Harvard University Press, 1990.

Brunswik, E. *Perception and the Representative Design of Psychological Experiments*. Berkeley: University of California Press, 1956.

Burchill, S., and A. Linklater. *Theories of International Relations*. New York: St. Martin's Press, 1996.

Burghardt, R. F., D. Jones, P. Clapp, J. Larocco, M. McAfee. *Who Needs Embassies?* Washington, DC: Georgetown University, 1997.

Burkhart, G. E., and S. Older, *The Information Revolution in the Middle East and North Africa*. Santa Monica, CA: RAND, 2003.

Burrell, G., and G. Morgan. *Sociological Paradigms and Organizational Analysis*. London: Heinemann, 1979.

Burt, R., O. Robison, and B. Fulton. *Reinventing Diplomacy in the Information Age*. Center for Strategic and International Studies. October 9, 1998.

Burton, Jr., D. F. "The Brave New Wired World." *Foreign Policy* (Spring 1997).

Bush, G. W. *Defending America's Cyberspace—National Plan for Information Systems Protection 1.0—An Invitation to Dialogue*. Washington, DC: Executive Office of the President, January 2000.

Cairncross, F. *The Death of Distance*. Boston: Harvard Business School Press, 1997.

Campen, A. D., ed. *First Information War: The Story of Communications, Computers, and Intelligence Systems in the Persian Gulf*. Fairfax, VA: AFCEA International Press, 1992.

Campen, A. D., and D. H. Dearth, eds. *CyberWar 2.0: Myths, Mysteries and Reality*. Fairfax, VA: AFCEA International Press, 1998.

———. *CyberWar 3.0: Human Factors in Information Operations and Future Conflict*. Fairfax, VA: AFCEA International Press, 2000.

Campen, A. D., D. H. Dearth, and R. T. Goodden, eds. *CyberWar: Security, Strategy and Conflict in the Information Age*. Fairfax, VA: AFCEA International Press, 1996.

Capra, F. *The Web of Life: A New Synthesis of Mind and Matter*. London: Flamingo, 1996.

Carr, E. H. *The Twenty Years' Crisis, 1919–1939*. New York: Harpers Perennial, 1939.

Catton, J. Interview with author. June 10, 2003.

Cawkell, A. E., ed. *Evolution of An Information Society*. London: ASLIB, 1987.

Cebrowski, A. Interview with author. June 10, 2003.

Charmaz, K. "Learning Grounded Theory: Rethinking Psychology." *Handbook of Qualitative Research*. Thousand Oaks, CA: Sage Publications, Inc., 1995.

———. "Grounded theory: Objectivist and constructivist methods." *Handbook of Qualitative Research*. Thousand Oaks, CA: Sage Publications, Inc., 2000.

Checkland, P. B. *Systems Thinking, Systems Practice*. Chichester, UK: Wiley, 1981.

Checkland, P. B., and L. J. Davies. "The Use of the Term Weltanschauung is Soft Systems Management." *Journal of Applied Systems Analysis* 13 (1985): 109.

Checkland, P. B., and S. Howell. "Information Management and Organization Processes: An Approach through Soft Systems Management." *Journal of Information Systems* 1, Vol. 3 (1993): 3–16.

———. *Information, Systems and Information Systems*. Chichester, UK: Wiley, 1998.

Checkland, P. B., and J. Shoals. "Techniques of Soft Systems Practice, Part 4, Conceptual Model Building Revisited." *Journal of Applied Systems Analysis* 2, Vol. 17 (1989): 39.

———. *Soft Systems Management in Action*. Chichester, UK: Wiley, 1990.

Cherryholmes, C. H. "Notes on Pragmatism and Scientific Realism." *Educational Researcher* 6, Vol. 21 (August–September 1992): 13–17.

60 Minutes. Interview with Richard Clarke. CBS, April 9, 2000.

Clark, W. K. *Waging Modern War: Bosnia, Kosovo, and the Future of Combat*. New York: PublicAffairs, 2001.

Clarke, S. *Social Theory, Psychoanalysis, and Racism*. New York: Palgrave Macmillan, 2003.

Clodfelter, M. *The Limits of Air Power, The American Bombing of North Vietnam*. New York: Free Press, 1989.

Cloud, J. "The Manhunt Goes Global." *TIME* (October 2001): 52–57.

Cohen, E. "A Revolution in Warfare." *Foreign Affairs* 75 (March/April 1996): 37–54.

Copeland, T. E., ed. *The Information Revolution and National Security*. Carlisle, PA: Strategic Studies Institute, U.S. Army War College, 2000.

Cordesman, A. H. *Transnational Threats From the Middle East: Crying Wolf or Crying Havoc?* Carlisle Barracks, PA: Strategic Studies Institute, U.S. Army War College, 1999.

Cordesman, A. H., and J. G. Cordesman. *CyberThreats, Information Warfare and*

Critical Infrastructure Protection: Defending the U.S. Homeland. Westport, CT: Praeger, 2002.

Creswell, J. *Qualitative Inquiry and Research Design: Choosing Among Five Traditions.* Thousand Oaks: Sage Publications, Inc., 1998.

———. *Research Design: Qualitative, Quantitative, and Mixed Methods Approaches.* Thousand Oaks, CA: Sage Publications, Inc., 2003.

Creveld, M. "Die USA im Psychokrieg," *Der Welt.* March 3, 2002.

Crotty, M. *The Foundations of Social Research: Meaning and Perspective in the Research Process.* London: Sage Publications, Inc., 1998.

"Dawn of Information Age Will Change Military More Than Cold War's End." *Aerospace Daily* 174 (May 30, 1995): 325.

Dearth, D. Interview with author. June 6, 2001.

———. "Shaping the Information Space." *Journal of Information Warfare* (May 2002).

De Caro, C. "Operationalizing SOFTWAR." In *Cyberwar 2.0: Myths, Mysteries and Reality.* Edited by A. D. Campen and D. H. Dearth. Fairfax, VA: AFCEA International Press, 1998.

———. Interview with author. April 14, 2003; April 26, 2004.

Delly, P. Interview with author. June 1, 2001.

Denning, D. E. *Information Warfare and Security.* Reading, MA: Addison–Wesley, 1999.

———. Interview with author. February 19, 2003.

Denzin, N. *Interpretative Biography.* London: Sage Publications, Inc., 1989.

Der Derian, J. "The Space of International Relations: Simulation, Surveillance, and Speed." *International Studies Quarterly* 3, Vol. 34 (September 1990).

Dessler, D. "What's at Stake in the Agent-Structure Debate?" *International Organization* 3, Vol. 43 (Summer 1989).

Devereux, T. *Messenger Gods of Battle: Radio, RADAR, SONAR—The Story of Electronics in War.* London: Brassey's Inc., 1991.

Devost, M. Interview with author. July 2, 2003.

Dominguez, R. Interview with author. September 10, 1999.

Dorflein, C. Interview with author. May 12, 2000.

Downs, G., ed. *Collective Security Beyond the Cold War.* Ann Arbor: University of Michigan Press, 1994.

Doyle, M. W. "Liberalism and World Politics." *American Political Science Review* 80 (December 1986): 1151–1169.

Duffy, M. "War on All Fronts." *TIME* (October 15, 2001): 30–37.

Dunn, M. *Information Age Conflicts: A Study of the Information Revolution and a Changing Operating Environment*. Zurich: Swiss Federal Institute of Technology, 2002.

Easterby-Smith, M., R. Thorpe, and A. Lowe. *Management Research*. London: Sage Publications, Inc., 1993.

Edwards, M. "The Foreign Affairs Budget." *The Brookings Review* (Spring 1997).

Edwards, P. N. *The Closed World: Computers and the Politics of Discourse in Cold War America*. Cambridge, MA: MIT Press, 1996.

Ehteshami, A. "The Arab States and the Middle East Balance of Power." In *Iraq, the Gulf Conflict and the World Community*. Edited by James Gow, 65. London: Brassey's Inc., 1993.

"Eight Down, Many More to Go." *The Economist* (November 22, 2001): 18.

Erlandson, D. A. *Doing Naturalistic Inquiry: A Guide to Methods*. London: Sage Publications, Inc., 1993.

Esterline, J. H., and Mae H. Esterline. *Innocents Abroad: How We Won the Cold War*. Lanham, MD: University Press of America, 1997.

Evera, S. "Offense, Defense, and the Causes of War." In *International Security* 4, Vol. 22 (Fall 1998): 5–43.

Felman, M. "The Military/Media Clash and the New Principle of War, Media Spin." Thesis presented to the School of Advanced Air Power Studies, Maxwell AFB, AL, 1992.

Finnemore, M. "Norms, Culture, and World Politics: Insights from Sociology's Institutionalism." *International Organization* 2, Vol. 50 (Spring 1996).

Fisher, G. *Mindsets: The Role of Culture and Perception in International Relations*. Yarmouth, MA: Intercultural Press, 1997.

———. *Public Diplomacy and the Behavioral Sciences*. Bloomington: Indiana University Press, 1972.

Flanagan, S. J., E. L. Frost, and R. L. Kugler. *Challenges of the Global Century: Report of the Project on Globalization and National Security*. Washington, DC: Institute for National Strategic Studies, National Defense University, 2001.

Fletcher School of Law and Diplomacy. "The Future of Public Diplomacy." *Occasional Papers in Leadership and Statecraft* (March 1998).

Flinders, D. J., and G. E. Mills. *Theory and Concepts in Qualitative Research*. New York: Teachers College Press, 1993.

Flood, R. L., and M. C. Jackson. *Creative Problem Solving: Total Systems Intervention.* London: John Wiley & Sons, 1991.

Foer, F. "Flacks Americana—John Rendon's Shallow PR War on Terrorism." *New Republic* (May 20, 2002).

Forno, R. Interview with author. April 21, 2003.

Forno, R., and R. Baklarz. *The Art of Information Warfare.* Washington, DC: Universal Publishers, 1999.

Fukuyama, F., and A. N. Shulsky. "Military Organization in the Information Age: Lessons From the World of Business." In *Strategic Appraisal: The Changing Role of Information in Warfare.* Edited by Zalmay Khalilzad and John White, 327–360. Santa Monica, CA: RAND, 1999.

Fukuyama, F., and C. Wagner. *Information and Biological Revolutions: Global Governance Challenges.* Santa Monica, CA: RAND, 2000.

Fulton, B. "Learning from Business—New, 'Improved' State Should be Creative, Dynamic, Decentralized." *Foreign Service Journal* (December 1997).

———. *Reinventing Diplomacy in the Information Age.* Washington, DC: Center for Strategic and International Studies, 1998.

———. Interview with author. June 10, 2003.

Galbraith, P. "The Decline and Fall of USIA." *Foreign Service Journal* (September 1999).

Ganley, O. H., and G. D. Ganley. *To Inform or To Control? The New Communications Networks.* Second Edition. Norwood, NJ: Ablex Publishing Corp., 1989.

Geva, N., and A. Mintz, eds. *Decision-Making on War and Peace: The Cognitive-Rational Debate.* Boulder, CO: Lynne Rienner Publishers, 1997.

Gibson, J. *The Ecological Approach to Perception.* Boston: Houghton Mifflin, 1979

Giessler, F. Interview with author. April 17, 2003; March 25, 2004.

Glaser, B. G., and A. Strauss. *The Discovery of Grounded Theory: Strategies for Qualitative Research.* New York: Adine de Gruyter, 1967.

Glaser, C. L. "Realists as Optimists: Cooperation as Self-Help." *International Security* 3, Vol. 19 (September 2009): 50–90.

Gleick, J. *Faster.* New York: Vintage Books, 1999.

———. *What Just Happened: A Chronicle from the Information Frontier.* New York: Pantheon Books, 2002.

Goggett, C. L., and L. T. Goggett. *The United States Information Agency.* New York: Chlesea House, 1989.

Gompert, D. C. *Right Makes Might: Freedom and Power in the Information Age*. Washington, DC: National Defense University, 1998.

————. Interview with author. April 21, 2003.

Gompert, D. C., R. L. Kugler, and M. C. Libicki. *Mind the Gap: Promoting a Transatlantic Revolution in Military Affairs*. Washington, DC: Institute for National Strategic Studies, National Defense University, 1999.

Gordon, M. R., and B. E. Trainor. *The Generals' War: The Inside Story of the Conflict in the Gulf*. Boston: Little, Brown and Company, 1995.

Gow, J., ed. *Iraq, the Gulf Conflict and the World Community*. London: Brassey's Inc., 1993.

Gowing, N. "Media Coverage: Help or Hindrance in Conflict Prevention?" *Carnegie Commission on Preventing Deadly Conflict*, Prepublication Draft (May 1997).

Graham, B. "Bush Orders Guidelines for Cyber-Warfare Rules for Attacking Enemy Computers Prepared as U.S. Weighs Iraq Options." *Washington Post*. February 7, 2003.

Gray, C. H. *Postmodern War: The New Politics of Conflict*. New York: The Guilford Press, 1997.

Greenberg, L. T., S. E. Goodman, and K. J. Soo Hoo. *Information Warfare and International Law*. Washington, DC: Institute for National Strategic Studies, National Defense University, 1997.

Gregory, B. Interview with author. April 23, 2003.

Guba, E. G. *Curriculum Inquiry* 1, Vol. 22 (Spring, 1992): 17–23.

Guba, E., and Y. Lincoln. "Do inquiry paradigms imply inquiry methodologies?" In *Qualitative Approaches to Evaluation in Education*. Edited by D. M. Fetterman. Westport, CT: Praeger, 1988.

Haave, K. Interview with author. April 15, 2003.

Habermas, J. *Theory and Practice*. London: Heinemann, 1974.

Hachigan, N. "China's Cyber Strategy." *Foreign Affairs* (March / April 2001).

Hall, W. M. *Stray Voltage: War in the Information Age*. Annapolis, MD: Naval Institute Press, 2003.

Hamblet, W. P., and J. G. Kline. "Interagency Cooperation: PDD 56 and Complex Contingency Operations." *Joint Forces Quarterly* (Spring 2000): 92–97.

Hansen, A. C. *USIA, Public Diplomacy in the Computer Age*. Second Edition. Westport, CT: Praeger, 1989.

Harnard, S. "Neoconstructivism: A Unifying Theme for the Constructive Sciences."

In *Language, Mind and Brain*. Edited by T. Simons and R. Scholes. Hillsdale, NJ: Erlbaum, 1982.

Haskell, B. H. "Access to Society: A Neglected Dimension of Power." *International Organization* 34 (Winter 1980): 89–120.

Hawley, L. Interview with author. May 1, 2000.

Hayes, M. D., and G. F. Wheatley, eds. *Interagency and Political-Military Dimensions of Peace Operations: Haiti-A Case Study.* Washington, DC: Institute for National Strategic Studies, National Defense University, 1996.

Henry, R., and E. C. Peartree, eds. *The Information Revolution and International Security.* Washington, DC: Center for Strategic and International Studies Press, 1998.

Hoffman, D. "Beyond Public Diplomacy." *Foreign Affairs* (March/April 2002).

Holsti, O. "Theories of International Relations and Foreign Policy: Realism and its Challengers." In *Controversies in International Relations Theory: Realism and the Neoliberal Challenge.* Edited by Charles W. Kegley. New York: St Martin's Press, 1995.

Honderich, T. *The Oxford Companion to Philosophy.* Oxford: Oxford University Press, 1995.

Hormats, R. D. "Foreign Policy by Internet." *Washington Post.* July 29, 1997.

Hubbard, Z. Interview with author. August 12, 2003; November 19, 2003; April 23, 2004.

Hughes, J. "Don't Weaken USIA: War on Words Continues." *Wall Street Journal,* August 8, 1988.

Hundley, R. O., R. H. Anderson, T. K. Bikson, M. Botterman, J. Cave, C. R. Neu, M. Norgate, and R. Cordes. *The Future of the Information Revolution in Europe: Proceedings of an International Conference.* Santa Monica, CA: RAND, 2001.

Hundley, R. O., R. H. Anderson, T. K. Bikson, J. A. Dewar, J. Green, M. Libicki, and C. R. Neu. *The Global Course of the Information Revolution: Political, Economic, and Social Consequences—Proceedings of an International Conference.* Santa Monica, CA: RAND, 2000.

Hurd, D. "Has Diplomacy a Future?" *Ditchley Foundation Lecture XXXIII.* 1996.

Huth, P., and B. Russett. "What Makes Deterrence Work? Cases from 1900 to 1980." *World Politics* 4, Vol. 36 (July 1984): 496–526.

Jackson, M. *System Approaches to Management.* New York: Kluwer Academic/Plenum Publishers, 2000.

Jervis, R. *Perception and Misperception in International Politics.* Princeton: Princeton University Press, 1976.

Johnson, J. Interview with author. April 16, 2003; November 24, 2003; March 26, 2004.

Johnson, R. B., and A. J. Onwuegbuzie. "Mixed methods research: A research paradigm whose time has come." *Educational Researcher* 7, Vol 33 (2004): 14–26.

Johnson, S. E., and M. C. Libicki. *Dominent Battlespace Knowledge.* Washington, DC: Institute for National Strategic Studies, National Defense University, 1996.

Johnstone, C. "The End of Foreign Policy." *World Affairs and Diplomacy in the 21st Century.* Washington, DC: Dacor Bacon House Foundation Conference, 1997.

Jones, J. Interview with author. August 13, 2003; April 1, 2004.

Kast, D., and R. L. Kahn. *The Social Psychology of Organizations.* Chichester, UK: Wiley, 1966.

Kast, F. E., and J. E. Rosenzweig. *Organization and Management: A System and Contingency Approach.* New York: McGraw-Hill, 1981.

Kegley, C. W. "The Neoliberal Challenge to Realist Theories of World Politics: An Introduction." In *Controversies in International Relations Theory: Realism and the Neoliberal Challenge.* Edited by Charles W. Kegley. New York: St Martin's Press, 1995.

Keith, K. "Troubled Takeover: The Demise of USIA." *Foreign Service Journal* (September 1999).

Kemmis, S., and M. Wilkinson. "Participatory Action Research and the Study of Practice." In *Action Research in Practice: Partnerships for Social Justice in Education.* Edited by Bill Atweh, Stephen Kemmis, and Patricia Weeks. New York: Routledge, 1998.

Keohane, R. O. *After Hegemony: Cooperation and Discord in the World Political Economy.* Princeton, NJ: Princeton University Press, 1984.

Keohane, R. O., and J. S. Nye. *Power and Interdependence.* Second Edition. Boston: Longman, 1989.

———. "Power and Interdependence in the Information Age." *Foreign Affairs* 77 (September/October 1998): 81–94.

Kerlinger, F. *Behavioral Research: A Conceptual Approach.* New York: Holt, Rinehart, and Winston, 1979.

Khalilzad, Z., and J. White, eds. *Strategic Appraisal: The Changing Role of Information in Warfare.* Santa Monica, CA: RAND, 1999.

Kilroy, R. Interview with author. August 7, 2003.

Kopp, C. Interview with author. August 6, 2003.

Kovach, P. Interview with author. April 22, 2003; March 31, 2004.

Kuehl, D. "Defining Information Power." *Strategic Forum* 115 (June 1997).

———. Interview with author. July 4, 2003; April 1, 2004.

Kuehl, D., and R. Neilson. "No Strategy for the Information Age." *US Naval Institute Proceedings* (September 2003): 2.

Kuhn, T. *The Structure of Scientific Revolutions.* Chicago: University of Chicago Press, 1970.

Kuusisto, T., R. Kuusisto, and E. L. Armistead. "System Approach to Information Operations." *3rd European Conference on Information Warfare and Security* (2004), 231–239.

Lagana, G. Interview with author. April 24, 2003.

Laird, R. F., and H. Holger. *The Revolution in Military Affairs: Allied Perspectives.* Washington, DC: Institute for National Strategic Studies, National Defense University, 1999.

Lauer, M. Interview with author. June 10, 2003; April 1, 2004.

Lebow, R. N., and J. G. Stein. "Deterrence: The Elusive Dependent Variable." *World Politics* 3, Vol. 42 (April 1990).

Lennon, A. T. J., ed. *The Battle for Hearts and Minds: Using Soft Power to Undermine Terrorist Networks.* Cambridge, MA: MIT Press, 2003.

Leonard, D., and R. McAdam. "Grounded Theory Methodology and Practitioner Reflexivity in TQM Research." *International Journal of Quality and Reliability Management* 2, Vol. 18 (2001): 180–194.

Levien, F. H. "Kosovo: An IW Report Card." *Journal of Electronic Defense* (August 1999).

Libicki, M. C. *The Mesh and the Net: Speculations on Armed Conflict in a Time of Free Silicon.* Washington, DC: Institute for National Strategic Studies, National Defense University, August 1995.

———. *Standards: The Rough Road to the Common Byte.* Washington, DC: Institute for National Strategic Studies, National Defense University, 1995.

———. *What is Information Warfare?* Washington, DC: Institute for National Strategic Studies, National Defense University, 1995.

———. *Defending Cyberspace and Other Metaphors.* Washington, DC: Institute for National Strategic Studies, National Defense University, 1996.

———. *Who Runs What in the Global Information Grid: Ways to Share Local and Global Responsibility.* Santa Monica, CA: RAND, 2000.

———. Interview with author. April 22, 2003.

Lincoln, Y. S. "Curriculum Studies and Traditions of Inquiry: The Humanistic Tradition." In *Handbook of Research on Curriculum*. Edited by P. W. Jackson. New York: Macmillian, 1992.

Lincoln, Y. S., and E. G. Guba. *Naturalistic Inquiry*. London: Sage Publications, Inc., 1985.

Livingston, S. *Clarifying the CNN Effect: An Examination of Media Effects According to Type of Military Intervention.* Research Paper R-18, presented at the Joan Shorenstein Center on the Press, Politics and Public Policy, Harvard University, June 1997.

Livingston, S., and T. Eachus. "Humanitarian Crises and U.S. Foreign Policy: Somalia and the CNN Effect Reconsidered." *Political Communication* 12 (1995).

Loke, W. H., ed. *Perspectives on Judgement and Decision Making*. Lanham, MD: Scarecrow Press, 1996.

Lord, C. *The Past and Future of Public Diplomacy*. Greenwich, NY: ORBIS, 1998.

Malone, G., ed. *American Diplomacy in the Information Age*. New York: University Press of America, 1991.

Malone, J. Interview with author. July 4, 2003.

Marczyk, G., D. DeMatteo, and D. Festinger. *Essentials of Research Design and Methodology*. London: John Wiley & Sons, 2005.

Maoz, Z. *Paradoxes of War: On the Art of National Self-Entrapment*. Boston: Unwin Hyman, 1990.

McGeary, J. "Can the Afghans Come Together?" *TIME* (November 26, 2001): 48–49.

McNamara, D. Interview with author. April 22, 2003.

McPherson, M. Interview with author. June 5, 2000.

Mearsheimer, J. J. "Back to the Future: Instability in Europe after the Cold War." *International Security* 1, Vol. 15 (1990).

———. "A Realist Reply." *International Security* 1, Vol. 20 (Fall 1995): 92.

Melanson, R. A. *American Foreign Policy Since the Vietnam War: The Search for Consensus from Nixon to Clinton*. Armonk, NY: M. E. Sharpe, 2000.

Mermin, J. "Television News and American Intervention in Somalia: The Myth of a Media-Driven Foreign Policy." *Political Science Quarterly* 3, Vol. 112 (Fall 1997).

Merriam, S. B. *Case Study Research in Education: A Qualitative Approach*. San Francisco, CA: Jossey-Bass, Inc., 1988.

Metzl, J. "Rwandan Genocide and the International Law of Radio Jamming." *American Journal of International Law* 4, Vol. 91 (October 1997): 628–651.

———. "Network Diplomacy." *Georgetown Journal of International Affairs* I, Vol. II (Winter/Spring 2001).

———. Interview with author. April 16, 2003.

Miller, S. "International Security at Twenty-Five: From One World to Another." *International Security* 1 Vol. 26 (Summer 2001): 5–39.

Moisy, C. "Myths of the Global Information Village." *Foreign Policy* (Summer 1997).

Molander, R. C. Interview with author. May 13, 2003.

Molander, R. C., et al., eds. *Strategic Information Warfare Rising.* Santa Monica, CA: RAND, 1998.

Montgomery, M. Interview with author. June 5, 2000.

Morgenthau, H. J. *Politics Among Nations: The Struggle for Power and Peace.* New York: Alfred A. Knopf, 1967.

Morse, J. *Strategies for Sampling, Qualitative Nursing Research: A Contemporary Dialogue.* London: Sage Publications, Inc., 1991.

Moteff, J. D. *Critical Infrastructures: Background, Policy and Implementation.* Congressional Research Report RL30153. Washington, DC: Congressional Research Service, August 7, 2003.

Moustakas, C. *Phenomenological Research Methods.* Thousand Oaks, CA: Sage Publications, Inc., 1994.

Munro, N. *The Quick and the Dead: Electronic Combat and Modern Warfare.* New York: St Martin's Press, 1991.

———. "Inducting Information." *National Journal* (March 27, 1999): 818.

———. "Infowar: AK-47s, Lies and Videotape." *Association for Computing Machinery, Communications of the ACM* (July 1999).

Naef, W. Interview with author. July 4, 2003.

Nagel, E. "The Structure of Science." *Americal Journal of Physics* 10, Vol. 29 (October 1961).

National Communications System. *The Electronic Intrusion Threat to National Security and Emergency Preparedness Telecommunications: An Awareness Document.* Washington, DC: National Communications System, August 1993.

Nayar, B. R. "Regimes, Power and International Aviation." *International Organization* 1, Vol. 49 (Winter, 1995).

Neilson, R. E. ed. *Sun Tzu and Information Warfare.* Washington, DC: Institute for National Strategic Studies, National Defense University, 1997.

Neuman, J. *Lights, Camera, War: Is Media Technology Driving International Politics?* New York: St. Martin's Press, 1996.

Newsom, D. *The Public Dimension of Foreign Policy*. Bloomington: Indiana University Press, 1996.

Nicander, L. Interview with author. July 3, 2003.

Nye, Joseph S. *Bound to Lead: The Changing Nature of American Power*. New York: Basic Books, Inc., 1990.

———. "Redefining the National Interest." *Foreign Affairs* 78 (July/Aug 1999): 22–35.

———. *Power in the Global Information Age: From Realism to Globalization*. New York: Routledge, 2004.

Nye, J. S., and W. A. Owens. "America's Information Edge." *Foreign Affairs* 75 (March/April 1996): 20–36.

O'Neil, R. Interview with author. April 16, 2003.

Owen, J. M. "How Liberalism Produces Democratic Peace." *International Security* 2, Vol. 19 (1998): 87–125.

Owens, W., and E. Offley. *Lifting the Fog of War*. Baltimore, MD: The John Hopkins University Press, 2001.

Pachios, H. C. "Plain English for Public Diplomacy." *Washington Times*. August 17, 1999.

———. "A New Direction for the State Department." *Boston Globe*. February 26, 2001.

Parker, W. Interview with author. August 13, 2003; March 24, 2004.

Patton, M. Q. *Qualitative Evaluation and Research Methods*. Second Edition. Newbury Park, CA: Sage Publications, Inc., 1990.

Pendergrast, D. "State and USIA: Blending a Dysfunctional Family." *Public Diplomacy Foundation*. December 1999.

Peterson, S. *Crisis Bargaining and the State: The Domestic Politics of International Conflict*. Ann Arbor: The University of Michigan Press, 1996.

Phillips, D. C., and N. C. Burbules. *Postpositivism and Educational Research*. Lanham, MD: Rowman and Littlefield, 2000.

Pickard, A. J. *Access to electronic information resources: their role in the provision of learning opportunities to young people. A constructivist inquiry*. Unpublished doctoral dissertation. Newcastle upon Tyne, UK: Northumbria University, 2002.

Pickard, A. J., and D. P. Pickard "The applicability of constructivist user studies: How can constructivist inquiry inform service providers and systems designers?" *Information Research* 3, Vol. 9 (March 2004).

Pilecki, C. Interview with author. June 6, 2000; April 16, 2002.

Polkinghorne, D. E. "Phenomenological Research Methods." In *Existential-Phenom-*

enological Perspectives in Psychology. Edited by R. S. Valle and S. Halling, 3–16. New York: Plenum, 1989.

Pound, E. "The Root of All Evil." *U.S. News and World Report* (December 3, 2001): 29–30.

Radvanyi, J. ed. *Psychological Operations and Political Warfare in Long-Term Strategic Planning.* Westport, CT: Praeger, 1990.

Rattray, G. *Strategic Warfare in Cyberspace.* Cambridge, MA: MIT Press, 2001.

Ray, J. L. "Promise or Peril? Neorealism, Neoliberalism, and the Future of International Politics." In *Controversies in International Relations Theory: Realism and the Neoliberal Challenge.* Edited by Charles W. Kegley. New York: St Martin's Press, 1995.

Reid, R. P. "Waging Public Relations: A Cornerstone of Fourth-Generation Warfare." *Journal of Information Warfare* (May 2002).

Rendon, J. Interview with author. April 18, 2003; April 1, 2004.

Rew, L., D. Bechtel, and A. Sapp, "Self as an Instrument in Qualitative Research." *Nursing Research* 5, Vol. 42 (September/October 1993): 300–301.

Rheingold, H. *The Virtual Community.* Cambridge, MA: MIT Press, 2000.

———. *Smart Mobs: The Next Social Revolution.* Cambridge, MA: Perseus Publishing, 2003.

Ricks, T. "Downing Resigns as Bush Aide." *Washington Post.* June 28, 2002.

Riegert, K. "Know Your Enemy, Know Your Allies: Lessons Not Learned from the Kosovo Conflict." *Journal of Information Warfare* (May 2002).

Ronfeldt, D., J. Arquilla, G. E. Fuller, and M. Fuller. *The Zapatista Social Netwar in Mexico.* Santa Monica, CA: RAND, 1998.

Rosecrance, R. N. *The Rise of the Virtual State: Wealth and Power in the coming Century.* New York: Basic Books, Inc., 1999.

Rosenau, J. S. "Global Affairs in an Epochal Transformation." In *The Information Revolution and International Security.* Edited by Ryan Henry and C. Edward Peartree, 33. Washington, DC: Center for Strategic and International Studies, 1998.

Rötzer, F. "Aus für die Propaganda-Abteilung des Pentagon" [From the Propaganda Department of the Pentagon], *Die Heise.* February 21, 2002.

———. "Rumsfeld: Pentagon Lügt Nicht" [Rumsfeld: The Pentagon Does Not Lie], *Die Heise.* February 21, 2002.

———. "Schon Wieder Eine Chinesische Mai-Offensive?" [Another Chinese Offensive in May?] *Die Heise.* April 25, 2002.

Rowlands, T. Interview with author. July 2, 2003.

Rumsfeld, D. U.S. Congress, House Armed Services Committee. *Prepared Statements by Secretary of Defense in the FY 2004 Defense Budget Hearings.* February 5, 2003.

Scarborough, R. "Study hits White House on Peacekeeping Missions." *Washington Times.* December 6, 1999.

Schon, D. A. *The Reflective Practitioner: How Professionals Think in Action.* New York: Basic Books, Inc., 1983.

Schwartau, W. *Information Warfare: Cyberterrorism—Protecting Your Personal Security in the Electronic Age.* New York: Thunder's Mouth Press, 1996.

———. Interview with author. July 3, 2003.

Schwartzstein, S. J. D., ed. *The Information Revolution and National Security Dimensions and Directions.* Washington, DC: Center for Strategic and International Studies, 1996.

Science Applications International Corporation. *Information Warfare: Legal, Regulatory, Policy and Organizational Considerations for Assurance.* Report to the Joint Staff, Science Applications International Corporation, McLean, Virginia. July 4, 1995.

Seale, C., ed. *Researching Society and Culture.* London: Sage Publications, Inc., 1998.

Selznick, P. "Foundations of the Theory of Organization." *American Sociological Review* 1, Vol. 13 (February 1948): 25–35.

Senge, P. M. *The Fifth Discipline: The Art and Practice of the Learning Organization.* London: Random House, 1990.

Shapiro, H. "Learning from Failure—FSO Says Agency Has 'Implausible Goals & Wastes Time, Money.'" *Foreign Service Journal* (December 1997).

Shapiro, M. J. "Strategic Discourse/Discursive Strategy: The Representation of 'Security Policy' in the Video Age." *International Studies Quarterly* 3, Vol. 34 (September 1990).

Shultz, G. *Public Diplomacy in the Information Age.* Washington, DC: U.S. Department of State Bulletin, November 1987.

Sieber, S. D. "The Integration of Fieldwork and Survey Methods." *American Journal of Sociology* 6, Vol. 78 (May 1973): 1335–1359.

Siegel, P. *Target Bosnia: Integrating Information Activities in Peace Operations: NATO-Led Operations in Bosnia-Herzegoviona, December 1995–1997.* Washington, DC: National Defense University, 1998.

———. Interview with author. April 17, 2003; March 25, 2004.

Simmons, B. *Who Adjusts? Domestic Sources of Foreign Economic Policy during the Interwar Years 1923–1939*. Princeton, NJ: Princeton University Press, 1994.

Simpson, C. *Science of Coercion: Communication Research and Psychological Warfare, 1945–1960*. Oxford: Oxford University Press, 1994.

Snow, N. *Propaganda, Inc.: Selling America's Culture to the World*. New York: Seven Stories Press, 1998.

———. *Information War: American Propaganda, Free Speech and Opinion Control since 9-11*. New York: Seven Stories Press, 2003.

Stagg, V. Interview with author, July 4, 2003.

Stake, R. *The Art of Case Research*. Thousand Oaks, CA: Sage Publications, Inc., 1995.

Steele, R. D. *Information Operations—All Information, All Languages, All the Time: The New Semantics of War & Peace, Wealth & Democracy*. Oakton, VA: Oakton Press, 2006.

Stern, P. N. "Grounded Theory Methodology: Its Uses and Processes." *Image* 12 (1980) 20–23.

Stoll, C. *The Cuckoo's Egg: Tracking a Spy Through the Maze of Computer Espionage*. New York: Pocket Books, 1990.

Strauss, A., and J. Corbin. *Basics of Qualitative Research: Techniques and Procedures for Developing Grounded Theory*. Thousand Oaks, CA: Sage Publications, Inc., 1998.

Strobel, W. P. *Late-Breaking Foreign Policy: The News Media's Influence on Peace Operations*. Washington, DC: U.S. Institute of Peace Press, 1997.

Summe, J. Interview with author. June 10, 2003.

Sylvester, C. *Feminist Theory and International Relations in a Postmodern Era*. Cambridge: Cambridge University Press, 1994.

Taylor, Phillip. *War and the Media: Propaganda and Persuasion during the Gulf War*. Manchester, UK: Manchester University Press, 1992.

———. *Munitions of the Mind: A History of Propaganda from the Ancient World to the Present Day*. Manchester, UK: Manchester University Press, 1995.

———. *Global Communications, International Affairs and the Media since 1945*. New York: Routledge, 1997.

———. "Perception Management and the 'War' against Terrorism." *Journal of Information Warfare* (May 2002).

———. Interview with author. July 3, 2003.

Tellis, A. J., J. L. Bially, C. Layne, M. McPherson, and J. Sollinger. *Measuring National Power in the Postindustrial Age: Analyst's Handbook*. Santa Monica, CA: RAND, 2000.

Tempestilli, M. *Waging Information Warfare: Making the Connection Between Information and Power in a Transformed World.* Newport, RI: Naval War College, 1995.

Terriff, T., S. Croft, L. James, and P. M. Morgan. *Security Studies Today.* Cambridge, UK: Polity Press, 1999.

Tesch, R. *Qualitative Research: Analysis Types and Software Tools.* New York: Falmer, 1990.

Thomas, J. *Doing Critical Ethnography.* Newbury Park, CA: Sage Publications, Inc., 1993.

Thomas, T. *Cyber Silhouettes: Shadows over Information Operations.* Fort Leavenworth, KS: Foreign Military Studies Office, 2005.

Thompson, W. R., and L. Vescera. "Growth Waves, Systemic Openness, and Protectionism." *International Organization* 2, Vol. 46 (Spring 1992).

Timmes, T. Interview with author. June 12, 2001; April 15, 2002.

Toffler, A. *Future Shock.* New York: Bantam Books, 1970

———. *The Third Wave.* New York: Bantam Books, 1984.

Toffler, A., and H. Toffler. *War and Anti-War.* New York: Warner Books, 1993.

Torres, H. *Management Meaning: The Role of Psychological Operations and Public Diplomacy in a National Information Warfare Strategy.* Monterey, CA: Naval Postgraduate School, December 1995.

Treverton, G. F., and L. Mizell. *The Future of the Information Revolution in Latin America: Proceedings of an International Conference.* Santa Monica, CA: RAND, 2001.

Treverton, G. F., and S. G. Jones. *Measuring National Power.* Santa Monica, CA: RAND, 2005.

Ulhman, H. Interview with author. April 21, 2003.

Ullman, S. "Against Direct Perception." *Behavioral and Brain Sciences* 3, Vol. 3 (1980).

Ulrich, W. *Critical Heuristics and Social Planning: A New Approach to Practical Philosophy.* Chichester, UK: Wiley, 1995.

U.S. Advisory Commission on Public Diplomacy. *A New Diplomacy for the Information Age.* Washington, DC: State Department, 2000.

U.S. Commission on National Security/21st Century. *Roadmap for National Security: Imperative for Change.* February 15, 2001.

U.S. Department of the Air Force. Air Force Doctrine Document (AFDD) 2–5, *Information Operations* (January 2005).

U.S. Department of the Army. *Field Manual (FM) 3-13 Information Operations* (November 2003).

U.S. Department of Commerce. Information Infrastructure Task Force. *NII Security: The Federal Role*—Draft Report (June 5, 1995).

U.S. Department of Defense. Chairman, Joint Chiefs of Staff Instruction (CJCSI). *Joint Information Operations Policy* (U) S3210.01A (November 5, 1998).

U.S. Department of Defense. Chairman of the Joint Chiefs of Staff Memorandum of Policy 30 (CJCS MOP 30), *Command and Control Warfare* (March 8, 1993).

U.S. Department of Defense, DEPSECDEF MEMORANDUM, *Strategic Concept for Information Operations* (May 14, 1999).

U.S. Department of Defense Directive S3600.1, *Information Operations* (December 9, 1996).

U.S. Department of Defense Directive O-3600.1, *Information Operations* (August 14, 2006).

U.S. Department of Defense Directive 5200.28, *Security Requirements for Automated Information Systems* (AISs) (March 21, 1988).

U.S. Department of Defense Instruction. *Department of Defense Information Assurance Certification and Accreditation Process* (2007).

U.S. Department of Defense Instruction 5200.40. *DOD Information Technology Security Certification and Accreditation Process* (December 30, 1997).

U.S. Department of Defense. *Information Operations Roadmap* (2003).

U.S. Department of Defense. Joint Chiefs of Staff. Joint Publication 1-02, *Department of Defense Dictionary of Military and Associated Terms* (June 10, 1998).

U.S. Department of Defense. Joint Chiefs of Staff. Joint Publication 1-02, *Department of Defense Dictionary of Military and Associated Terms* (August 14, 2002).

U.S. Department of Defense. Joint Chiefs of Staff. *Information Assurance: Legal, Regulatory, Policy and Organizational Considerations* (August 1999).

U.S. Department of Defense. Joint Chiefs of Staff. Joint Publication 3-13, *Joint Doctrine for Information Operations* (October 9, 1998).

U.S. Department of Defense. Joint Chiefs of Staff. Joint Publication 3-13, *Joint Doctrine for Information Operations* (February 2006).

U.S. Department of Defense. Joint Chiefs of Staff, *Joint Vision 2010*. Washington, D.C., Government Printing Office (July 1996).

U.S. Department of Defense. Joint Chiefs of Staff. *Unified Command Plan Changes* (1999).

U.S. Department of Defense. *National Military Strategy to Secure Cyberspace* (2006).

U.S. Department of Defense. National Security Agency. Committee for National

Security Standards, No. 4012. *National Information Assurance Training Standard for Senior System Managers* (June 2004).

U.S. Department of Defense. National Security Agency. National Security Telecommunications and Information Systems Security, No. 4011. *National Training Standard for Information Systems Security (INFOSEC) Professionals* (June 20, 1994).

U.S. Department of Defense. National Security Agency. National Security Telecommunications and Information Systems Security, No. 4013. *National Training Standard for System Administrators (SA)* (March 2004).

U.S. Department of Defense. National Security Agency. National Security Telecommunications and Information Systems Security. *National Training Standard for Information Systems Security Officers* (April 2004).

U.S. Department of Defense. National Security Agency. National Security Telecommunications and Information Systems Security, No. 4015. *National Training Standard for System Certifiers* (December 2000).

U.S. Department of Defense. Office of the Under Secretary of Defense for Acquisition and Technology. *Report of the Defense Science Board Task Force on Information Warfare—Defense* (November 1996).

U.S. Department of Defense. Office of the Under Secretary of Defense for Acquisition, Technology & Logistics. *Report of the Defense Science Board Task Force on Managed Information Dissemination* (October 2001).

U.S. Department of Defense. *Quadrennial Defense Review Report* (September 30, 2001).

U.S. Department of Defense and Central Intelligence Agency. Memorandum of Agreement. *Establishment of Information Operation Technology Center* (March 4, 1997).

U.S. Department of Homeland Security. Homeland Security Presidential Directive 7, "Critical Infrastructure Identification, Prioritization, and Protection" (2003).

U.S. Department of the Navy. *Information Assurance Certification and Accreditation (C&A) Publication: Introduction to Certification and Accreditation*, Vol. 1 (December 2000).

U.S. Department of the Navy. *Information Assurance Certification and Accreditation (C&A) Publication: Certification and Accreditation of Site, Installed Program of Record, and Locally Acquired Systems*, Vol. 2 (December 2000).

U.S. Department of the Navy. *Information Assurance Certification and Accreditation (C&A) Publication: Certification and Accreditation of Program of Record Systems*, Vol. 3 (June 2000).

U.S. Department of the Navy. Chief of Naval Operations. *Information Assurance (IA) Program* (November 9, 1999).

U.S. Department of the Navy. *Marine Corps Warfighting Publication 3040.4, Marine Air-Ground Task Force Information Operations* (July 2003).

U.S. Department of State. *Consolidation of USIA into the State Department: An Assessment After One Year* (October 2000).

U.S. Department of State. *Foreign Affairs Reform and Restructuring Act H.R. 1757* (December 30, 1998).

U.S. Department of State. *National Strategy Plan to Govern US International Public Diplomacy and Strategic Communication.* Memo from Karen Hughes, Chair of the PCC on PD/SC to PCC Participants (October 18, 2006).

U.S. Department of State. *Reform and Restructuring Act* (December 30, 1998).

U.S. House of Representatives. Public Law—1757. Foreign Affairs Reform and Restructuring Act of 1998.

U.S. Information Agency. *United States Information Agency, A Commemoration* (September 1999).

U.S. President. Executive Order 12382. *National Security Telecommunications Advisory Committee* (September 13, 1982), continued by EO 12610 (September 30, 1987).

U.S. President. Executive Order 12472. *Assignment of National Security and Emergency Preparedness Telecommunications Functions* (April 3, 1984).

U.S. President. Executive Order 12881. *National Science and Technology Council* (November 23, 1993).

U.S. President. Executive Order 12882. *President's Committee of Advisers on Science and Technology* (November 23, 1993).

U.S. President. Executive Order 13010. *Presidential Commission on Critical Infrastructure Protection* (July 15, 1996).

U.S. President. Executive Order 13073. *President's Council on Year 2000 Conversion* (February 4, 1998).

U.S. President. Executive Order 13073. *President's Council on Year 2000 Conversion* (June 14, 1999).

U.S. President. Executive Order 13228. *Establishing the Office of Homeland Security and the Homeland Security Council* (October 8, 2001).

U.S. President. Executive Order 13231. *Critical Infrastructure Protection in the Information Age* (October 16, 2001).

U.S. President. Executive Order 13286. *Amendment of Executive Orders, and Other Actions, in Connection with the Transfer of Certain Functions to the Secretary of Homeland Security* (February 28, 2003).

U.S. President. *Interdepartmental Committee on Communications* (October 26, 1921). Updated (August 21, 1963).

U.S. President. National Security Council. *Coordination of Foreign Information Measures* 4 (December 9, 1947).

U.S. President. National Security Council Paper. *A Plan for National Psychological Warfare* 74 (July 10, 1950).

U.S. President. National Security Decision. *National Policy for the Security of National Security Telecommunications and Information Systems* 42 (July 5, 1990).

U.S. President. National Security Decision Directive. *US International Information Policy* 130 (March 6, 1984).

U.S. President. *National Security Presidential Directive.* NSPD-1 (February 23, 2001).

U.S. President. Executive Office of the President. *National Strategy for Combating Terrorism* (February 2003).

U.S. President. Executive Office of the President. *National Strategy for the Physical Protection of Critical Infrastructures and Key Assets* (February 2003).

U.S. President. National Security Telecommunications and Information Systems Security Committee Document. *Incident Response and Vulnerability Reporting for National Security Systems* 503 (August 30, 1993).

U.S. President. Office of Homeland Security. *National Strategy for Homeland Security* (July 16, 2002).

U.S. President. President's Commission on Critical Infrastructure Protection. *Critical Foundation: Protecting America's Infrastructures* (October 13, 1997).

U.S. President. Presidential Decision Directive 29, "Security Policy Coordination" (September 16, 1994).

U.S. President. Presidential Decision Directive 39, "US Policy on Counter-Terrorism" (June 21, 1995).

U.S. President. Presidential Decision Directive 56, "Managing Complex Contingency Operations" (May 1997).

U.S. President. Presidential Decision Directive 63, "Critical Infrastructure Protection" (May 22, 1998).

U.S. President. Presidential Decision Directive 68, "International Public Information" (April 30, 1999).

U.S. President. United States Commission on National Security/Twenty-First Century. *Roadmap for National Security: Imperative for Change* (February 15, 2001).

U.S. Senate. Central Intelligence Agency Director George J. Tenant before the United States Senate Select Committee on Intelligence (February 7, 2001).

U.S. Senate. *Report of the Commission on Protecting and Reducing Government Secrecy*. Senate Document 105-2 (1997).

Vatis, M. *Statement for the Record on Infrastructure Protection and the Role of the National Infrastructure Protection Center*. Presented before the Senate Judiciary Sub-Committee on Technology, Terrorism and Government Information. Washington, DC. June 10, 1998.

Von Bertalanffy, L. "The Theory of Open Systems im Physics and Biology." In *Systems Thinking*. Edited by F. E. Emery, Harmondsworth, UK: Penguin, 1950.

———. *General System Theory*. Harmondsworth, UK: Penguin, 1968.

Von Bulow, I. "The Bounding of a Problem Situation and the Concept of a System's Boundary in Soft Systems Methodology." *Journal of Applied Systems Analysis*, Vol. 16 (1989): 35–41.

Wallace, W. "Truth and Power, Monks and Technocrats: Theory and Practice in International Relations." *Review of International Studies* (July 1996): 311.

Waller, J. M. *Fighting the War of Ideas like a Real War*. Washington, DC: The Institute of World Politics Press, 2007.

———., ed. *The Public Diplomacy Reader*. Washington, DC: The Institute of World Politics Press, 2007.

Waltz, K. "Realist Thought and Neorealist Theory." *Journal of International Affairs* 44 (Spring/Summer 1990): 21–27.

Ward, B. Interview with author. June 2, 2001.

Weber, M. *The Theory of Social and Economic Organization*. New York: Free Press, 1964.

Webster, F. *Theories of the Information Society*. London: Routledge, 1995.

Wheatley, M. J. *Leadership and the New Science: Learning about Organization from an Orderly Universe*. San Francisco: Berrett-Koehler, 1992.

Whicker, M. L., J. P. Pfiffner, and R. A. Moore, eds. *The Presidency and the Persian Gulf War*. Westport, CT: Praeger, 1993.

Wick, C. Z. "The Future of Public Diplomacy." *Presidential Studies Quarterly* (Winter 1989).

Wiener, N. *Cybernetics*. Chichester, UK: Wiley, 1948.

Williamson, C. Interview with author. April 23, 2003; April 1, 2004.

Williamson, M., ed. *The Emerging Strategic Environment: Challenges of the Twenty-First Century*. Westport, CT: Praeger, 1999.

Wilson, B. *Systems: Concepts, Methodologies, and Applications*. New York: John Wiley & Sons, 1984.

————. *Soft Systems Methodology: Conceptual Model Building and its Contribution.* New York: John Wiley & Sons, 2001.

Wilson, P. Interview with author. April 25, 2003.

Woodward, B. *The Commanders.* New York: Simon and Schuster, 1991.

Woodward, B., and D. Balz. "Aus Trauer Wird Wut." *Der Welt.* February 16, 2002.

Wriston, W. R. "Bits, Bytes and Diplomacy." *Foreign Affairs* 76 (September/October 1997): 172–182.

Zacher, M. W., and R. A. Mathew. "Liberal International Theory: Common Threads, Divergent Strands." In *Controversies in International Relations Theory: Realism and the Neoliberal Challenge.* Edited by Charles W. Kegley. New York: St Martin's Press, 1995.

Index

About the Author

D r. Leigh Armistead is the director of business development for GbHawk LLC, the program chair for the International Conference of Information Warfare, an affiliate professor for the National Information Assurance Training and Education Center at Idaho State University, an adjunct lecturer at Edith Cowan University in Perth, Australia, and the Director of IO at Norwich University. He edits the *Journal of International Warfare* and is on the review board for the European Conference on Information Warfare. He has edited two volumes previously published by Potomac Books, Inc.: *Information Operations: Warfare and the Hard Reality of Soft Power* and *Information Warfare: Separating Hype from Reality*. He lives in Virginia Beach with his wife and their sons.